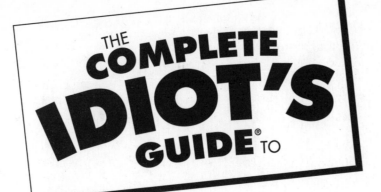

The Science of
Everything

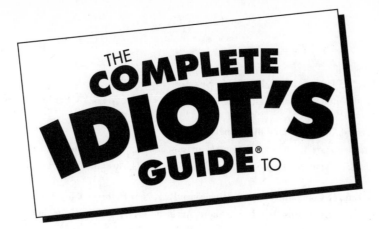

THE COMPLETE IDIOT'S GUIDE® TO

The Science of Everything

by Steve Miller

ALPHA

A member of Penguin Group (USA) Inc.

ALPHA BOOKS

Published by the Penguin Group

Penguin Group (USA) Inc., 375 Hudson Street, New York, New York 10014, USA

Penguin Group (Canada), 90 Eglinton Avenue East, Suite 700, Toronto, Ontario M4P 2Y3, Canada (a division of Pearson Penguin Canada Inc.)

Penguin Books Ltd., 80 Strand, London WC2R 0RL, England

Penguin Ireland, 25 St. Stephen's Green, Dublin 2, Ireland (a division of Penguin Books Ltd.)

Penguin Group (Australia), 250 Camberwell Road, Camberwell, Victoria 3124, Australia (a division of Pearson Australia Group Pty. Ltd.)

Penguin Books India Pvt. Ltd., 11 Community Centre, Panchsheel Park, New Delhi—110 017, India

Penguin Group (NZ), 67 Apollo Drive, Rosedale, North Shore, Auckland 1311, New Zealand (a division of Pearson New Zealand Ltd.)

Penguin Books (South Africa) (Pty.) Ltd., 24 Sturdee Avenue, Rosebank, Johannesburg 2196, South Africa

Penguin Books Ltd., Registered Offices: 80 Strand, London WC2R 0RL, England

International Standard Book Number: 978-1-59257-796-5
Library of Congress Catalog Card Number: 2008922783

10 09 08   8 7 6 5 4 3 2 1

Interpretation of the printing code: The rightmost number of the first series of numbers is the year of the book's printing; the rightmost number of the second series of numbers is the number of the book's printing. For example, a printing code of 08-1 shows that the first printing occurred in 2008.

Printed in the United States of America

Note: This publication contains the opinions and ideas of its author. It is intended to provide helpful and informative material on the subject matter covered. It is sold with the understanding that the author and publisher are not engaged in rendering professional services in the book. If the reader requires personal assistance or advice, a competent professional should be consulted.

The author and publisher specifically disclaim any responsibility for any liability, loss, or risk, personal or otherwise, which is incurred as a consequence, directly or indirectly, of the use and application of any of the contents of this book.

Most Alpha books are available at special quantity discounts for bulk purchases for sales promotions, premiums, fund-raising, or educational use. Special books, or book excerpts, can also be created to fit specific needs.

For details, write: Special Markets, Alpha Books, 375 Hudson Street, New York, NY 10014.

Publisher: *Marie Butler-Knight*
Editorial Director/Acquiring Editor: *Mike Sanders*
Senior Managing Editor: *Billy Fields*
Development Editor: *Michael Thomas*
Production Editor: *Kayla Dugger*
Copy Editor: *Amy Borelli*

Cartoonist: *Chris Sabatino*
Cover Designer: *Bill Thomas*
Book Designer: *Trina Wurst*
Indexer: *Johnna Vanhoose Dinse*
Layout: *Brian Massey*
Proofreaders: *Laura Caddell, Mary Hunt*

Contents at a Glance

Contents

Introduction

Did you ever wonder how somebody ever figured out all the details of making a cell-phone system work? What about those giant MRI machines that can take a picture of your insides without a single incision? Then there is the universe. We think we know at least something about how it started billions of years ago and where it's going billions of years in the future. How can we possibly know anything about an event that occurred 14 billion years ago?

Science is too often presented either as a huge database of facts or as an endeavor that is open only to a highly educated group of elite thinkers. It is neither of these. In its essence, science is a process, a way of looking at a question and finding an answer. Step-by-step, scientists look at a problem and ask, "How can I test this to see if it is true?" Asking one question at a time, they have built a body of knowledge too vast to be held in single library.

Although there is the occasional giant breakthrough, in which one person's insight changes the way we look at things, these are rare and momentous events—Newton and gravity, Darwin and natural selection, Einstein and relativity. And yet even Newton, whose insight stands out among the greatest of scientists, said "If I have seen farther, it is by standing on the shoulders of giants."

No one person set out to build a cell-phone system. It grew in tiny increments. Long ago, scientists found that radio waves could be transmitted and detected. Others figured out how to modulate those waves in order to reproduce a voice. Miniaturization of electronics with the invention of the transistor played a role. Circuit-printing techniques, battery technologies, digitized signals sorted by computers—each piece of the puzzle came from a different direction.

In this book, I have tried to show some of the background to science questions. To understand why the sky is blue, you have to know a little bit about light and something about the atmosphere. Once you know how the two things work together, the color to the sky makes sense. And it's also clear why the sky isn't always blue.

How This Book Is Organized

The book is organized in six parts. Part 1 is an overview of science, what it is, who does it, and why. This part also includes a brief summary of some of the big ideas, or theories, that drive scientific research today.

Parts 2 through 5 look at broad areas of science by asking and answering questions. The parts are broken down into physical sciences, biological sciences, earth and space sciences, and technology. In reality, these categories are just an organizing tool. Scientific research cannot be contained in such neat groups. Biology, physics, earth science, and medicine all overlap one another when the time comes to find the answer to a question.

Part 6 takes a look at where science has come from and where it is going. It covers some of the basic ideas that define how we see the world today, a few of the more well-known scientists among the vast number who have built our base of scientific knowledge, and a look at where the answers to today's questions may lead us. The appendix includes lists of places to look for more answers in science.

Despite the title, this is not really the science of everything. This book looks at a broad sampling of questions about science. Behind each answer is a lot of scientific research. In the end, the how and why of everything boils down to observation and understanding. I have tried to use these to answer the how and why.

Extras

There are some boxes with extra information here and there throughout the book to give a bit more information.

Ss Science Says

These are short quotes from scientists to get a look at how the people doing the research see science, life, or the world around them.

Us Uncommon Sense

Sometimes common sense and scientific understanding are very different. Sometimes new observations and discoveries make explanations obsolete. In either case, you may find that the why and how behind something are not at all what is commonly believed. These boxes point out some of those misconceptions.

Science includes a lot of jargon. Some of the terms that might be unfamiliar are defined in these boxes.

These are little bits of interesting information or observations related to the answer to a question.

Acknowledgments

The author wishes to thank Skip Press for his contributions to certain portions of this book.

Trademarks

All terms mentioned in this book that are known to be or are suspected of being trademarks or service marks have been appropriately capitalized. Alpha Books and Penguin Group (USA) Inc. cannot attest to the accuracy of this information. Use of a term in this book should not be regarded as affecting the validity of any trademark or service mark.

Part 1

Looking at the Universe Around Us

The basic assumption of all science is that nature operates by the same rules everywhere. Our lives are not driven by the whims of a supernatural world but by the clearly defined interactions of matter and energy. Why and how are questions that we pose to find those clear definitions.

In this part, we take a look at the role of science in our world and our lives. We also glance at some of the theories—explanations of the universe based on observation about it—that form the foundation of scientific research.

Chapter 1

What Is Science?

"Science is facts; just as houses are made of stones, so is science made of facts; but a pile of stones is not a house and a collection of facts is not necessarily science."

—*Henri Poincare (1854–1912)*

To many, science is nothing more than the isolation of a body of truths about a given subject—facts systematically gained through observation and experimentation. Science makes the world more easily understood and, for the most part, predictable.

More than just facts, though, science is a way of looking at the world. The scientific method is a process of questioning, observing, and developing explanations. Picture the cook in the kitchen, adjusting the spices in a pot of stew. The last time, the stew was a bit bland, so perhaps a little more pepper and a pinch of sage will produce just the right flavor this time. Or consider the farmer, planting corn in a field, trying several varieties to determine which gives the best yield for this particular type of soil. And, for that matter, watch a toddler playing in the sand at the beach, designing a structure that doesn't collapse. These are all examples of scientific research. They aren't funded by large research grants, and they don't result in a journal article, but each of these people is using the scientific method. Science is a systematic way of learning about the natural world.

Is science a natural human activity?

If trying to remember the difference between right-brain and left-brain activity leaves you yawning, just remember that some people are more interested in practicality and logic, while others are more given to creative and speculative activities. Good science requires a balance of the two. Science is a natural activity, because humans naturally attempt to discern how things work.

De ▐ Definition ▌

A **scientist** is by definition someone with an advanced knowledge of one or more sciences. It's really that simple, but keep in mind that the only scientists anyone remembers today, and the only ones anyone hires, are those who can apply that advanced knowledge in the real world.

Newborns feel attachment to parents long before they are born, and have no problem discerning the difference in their parents. That makes them little *scientists*, wouldn't you say? From learning words to taking her first steps, a baby learns by experimentation and mental notation of results.

People have been using science, often without realizing it, since the beginning of the human race. Early hunters learned how to make hunting tools by observation and experiment. If a particular type of stone was struck just so with the right tool, the resulting edge was sharp and lethal. Agriculture developed as gatherers noticed that specific seeds placed in the ground in the right way at the right season yielded more and better food than they found by chance.

Ff ▐ Fast Facts ▌

In a 2007 study detailed online in the journal *PLoS Computational Biology,* scientists stated that the brain can store more than 500 memories, but they could not be sure if memories were concentrated in one part of the brain, or spread across all the 100 billion brain cells making up the average adult brain. In the science of neuroscience, in which scientists using their brains try to understand our brains, they still aren't sure how it all works.

There's a Method in That Science

The modern scientific method was first laid out by Iraqi Arab Muslim scientist Ibn al-Haytham, also known as Alhacen, who chronicled his methods in his *Book of Optics* in 1021. (See Chapter 22 for more information on this scientist and his work.) This, combined with the syllogism (major premise, minor premise, and conclusion) developed by Greek philosopher Aristotle in his *Prior Analytics*, forms the basis of the current scientific method.

Alhacen's work influenced English philosopher and theologian Robert Grosseteste, whose commentary on Aristotle's *Posterior Analytics* included the idea that scientific reasoning should move from universal laws to particulars with predictable results. All these in turn influenced English Franciscan friar Roger Bacon, who was also known as Doctor Mirabilis ("wonderful teacher"). Bacon advised repeated observation of experimentation and a need for all results be independently verified. As such, all of his experiments were noted in such fine detail that anyone could independently reproduce and verify his results. Bacon was such a genius he was given a special compensation by Pope Clement IV allowing him to write on scientific matters.

Science did not become a formal discipline, however, until much later. During the Middle Ages, investigations that we would consider scientific today were described by the term "natural philosophy." Science, as a separate endeavor from philosophy, is a fairly recent idea. Science, as we know it today, means gaining knowledge through the scientific method.

The scientific method developed over the centuries, moving from philosophy and speculation to the necessity of verified repeatable results, and it is currently constituted of the following steps:

1. A question is presented.

2. The question is researched.

3. From the research a *hypothesis* is constructed.

4. The hypothesis is tested via experimentation.

> **De** | Definition
>
> A **hypothesis,** a word taken from the Greek *hypotithenai,* "to suppose," is a suggested explanation for any phenomenon, natural or otherwise. Hypotheses are different from theories in that they are, in the scientific method, either proven or disproven.

5. Data is analyzed and conclusions are drawn.

6. Results are communicated to the larger community.

7. Results are verified (or refuted) by others.

Whether determining the structure of DNA or discovering whether the planet Mars can sustain life, some form of the preceding steps is used by scientists of any discipline. Proving a hypothesis can take quite some time, however. In 1907, Albert Einstein in his theory of general relativity questioned Newtonian physics when he predicted that light would bend in a gravitational field. This was not proven until 1919, when astrophysicist Arthur Eddington made observations during a solar eclipse that verified Einstein's hypothesis.

Ss **Science Says**

"He that gives good advice, builds with one hand; he that gives good counsel and example, builds with both; but he that gives good admonition and bad example, builds with one hand and pulls down with the other."

—Francis Bacon (1561–1626)

Who does Science?

Imagine a place where people do scientific research. If I were to bet on what you are thinking of, I would place my money on a laboratory full of benches with strange equipment being operated by scientists in white lab coats. It's true that there are many laboratories like that in universities, hospitals, and manufacturing companies throughout the world. It is also true that a lot of scientific research occurs in these laboratories. Science, however, is not restricted to the stereotypical research facility.

In reality, everyone engages in science. You don't have to be a scientist; when you learn of a new idea or product and try it out, you are performing the last step of the scientific method. While everyone can look at a problem scientifically, the information in this book generally addresses science in its more common usage—the products of formal research by scientists. As stated in the title, science relates to everything around you in one way or another. However, by understanding and practicing the basic steps of the scientific method presented previously, more analytical results can be obtained in just about any area of life.

 Us   Uncommon Sense

It used to be that the label "Made in China" meant a product of lesser quality, but that has changed greatly in the twenty-first century, including in the field of science. According to the website www.sciencewatch.com, in March 2008 the People's Republic of China ranked number eight in the world as determined by scientific papers published in *Thomson Scientific-Indexed Journals of Physics* over the period 1997 to 2007. The United States was again ranked number one, but China's scientific prowess is now world-renowned.

Science as a Search for Knowledge

For the most part, all science comes about due to an effort to better the quality of life. From early tool making to the observations of heavenly events and their effects on crop production, the development of science has served to evolve humankind past mythology and superstition to predictability and improved conditions.

We still have a lot to learn. If we compared human civilization in recorded history to the normal education cycle in modern life, one could say that as a race we are barely out of high school. We have some idea how to deal with the weather, but rarely can we do anything about it. We have placed men on the moon, and had a robot take extensive pictures on Mars and transmit them back to Earth, but we are likely decades away from moving away from our home to other places where humans might live.

Ff   Fast Facts

Quantum entanglement is a scientific phenomenon in which two or more objects have a connection between each other that scientists do not fully understand. Also known as the "spooky effect," this interconnection is such that two photons could conceivably be separated across a galaxy, yet each "twin" would know what the other is doing. Experiments by French scientist Alain Aspect in the 1980s established this fairly conclusively, but there were loopholes that offered possible other explanations, so the jury is still out.

We've made a lot of progress in 10,000 years, though. Whereas science and philosophy or even religion were once inseparable in human thought, science can now be categorized in three areas: biological, physical, and social. Biological covers living things and their interaction, while the physical sciences deal with astronomy, chemistry, and all material things. The social sciences encompass everything related to behavior, including psychology and sociology.

Science and Technology

The wonderful thing about science is that its proven principles are demonstrable to anyone. When good science is developed, useful technology frequently results. One look at the consumer products resulting from the U.S. space program provides an excellent example. There's a long list at http://spaceplace.nasa.gov/en/kids/spinoffs2.shtml that includes the satellite dish, the ear thermometer, smoke detectors, ski boots, and joystick controllers.

These days, science and technology are inextricably woven together. If you would like to keep track of major developments, a good place to start is the magazine *Science & Technology Review*, published six times a year by the Lawrence Livermore Laboratory. For more information and to read articles like "Molecular Building Blocks Made of Diamonds," see www.llnl.gov/str.

Us **Uncommon Sense**

Alchemists in the Middle Ages vigorously pursued ways to turn lead into gold, which seems foolish today. Nevertheless, it really can be done, although not with a "philosopher's stone." The 1951 Nobel Laureate in Chemistry, Glenn Seaborg, transmuted a small quantity of lead into gold via physics in 1980. Eight years earlier, Soviet physicists at an experimental nuclear facility accidentally turned lead reactor shielding into gold. It can be done today in a particle accelerator, but the cost of the process is far more than the gold produced.

Everyone Can Understand Scientific Ideas

Given current challenges such as climate change and declining fishing yields, it is imperative that as many people as possible all over the world understand scientific ideas. That begins when people realize that science is not merely the province of "eggheads" but something learnable and applicable by anyone. If you understand the basic steps of the scientific method, you can "do science." If you understand basic scientific principles, you can invent items useful for everyone. Take the example of actress Hedy Lamarr, whose understanding of frequency hopping (rapid and random changing of the frequency of a radio signal) resulted in a patent intended to keep torpedoes from being thrown off course. Lamarr first learned of frequency hopping while sitting in on meetings with her Austrian munitions-dealer husband as arms designs were discussed. The ideas she patented are the basis for antijamming radio technology today.

The key to understanding science begins with the basics, and underlying the principles are the terms. In this book, you will find the terms defined wherever possible to their origins, so that the step-by-step developments of science are seen. By understanding the terms, you can understand the principles, and with the understanding of these basics you learn how to evaluate scientific evidence in general. Anyone can understand scientific ideas, but sometimes they first need to learn that such a thing is possible. Just as Hedy Lamarr gave rise to a major piece of technology by observation and experimentation, many great inventions have come about via the efforts of so-called average persons who were not afraid of delving into science.

What to Watch Out For

To make sure you are dealing with actual science, you need look no further than the Internet—not the use of the Internet, but the way it was initially put together. First developed in 1973 by Stanford graduate Vinton Cerf in conjunction with the United States Department of Defense Advanced Research Projects Agency (ARPA), the initial network was known as ARPAnet, and Cerf published the results of his work. ARPAnet linked computer networks at a few U.S. universities and research laboratories.

It was a decade before the network resembled anything that we know today. In 1989, the work was expanded into the World Wide Web by English computer scientist Tim Berners-Lee for the European Organization for Nuclear Research (CERN). By 1996, more than 25 million computers in over 180 countries were connected, and today it seems the entire world is accessible via the web. The lesson here is that scientific breakthroughs may arise from the single bright idea of one person or group of persons, but it often takes decades for the initial idea to be fully realized, and this is only after the initial results have been vetted and improved upon by other scientists.

Getting It Right and Getting It Wrong in the News

It may seem that the best source of information about science is the news media. In many ways this is true. Well-researched reports can convey information about science in a readable form. However, the media is not always completely reliable. One of the most famous incidents of misleading the public occurred on the radio when Orson Welles created a nationwide panic on October 30, 1938, with his presentation of H. G. Wells's *The War of the Worlds* on the CBS Radio Network. Welles presented the first part of the program as a series of news bulletins, and people across the United States thought the country was actually being invaded by Martians.

While this is an extreme example of getting the information wrong, major news outlets can—by selective editing, subject selection, or even opinion masked as news—be less than accurate about important facts. It is important to read, listen, and view critically. Most news reports will identify the researcher or institution responsible for the research behind the story. When it does not, it may be a good idea to maintain some skepticism, and be aware that the article could be pushing a viewpoint rather than reporting science.

Remember the steps of the scientific method as you evaluate science news. If the hypothesis has been thoroughly tested, the data well analyzed, and the published results verified by others, who are the others? A good reporter will investigate. What journalists know that you might not, however, is the idea that "if it bleeds, it leads," particularly with television news. If there is the possibility for scientific news to have a deeply emotional element, chances are it will, to attract the media. To view it scientifically, you need to strip away the emotion and look at the facts. That's what good scientists will do.

Ff **Fast Facts**

A great many products are advertised as "scientifically proven," but that can be a misnomer. Try finding a standard definition for that phrase; you probably can't. If you see that advertised, it would be wise to ask, "Whose science?"

Though there is science news, there is no organized science of newsgathering. The closest thing to it is computer-assisted reporting, which began with the rise of personal computers. It involves database and statistical analysis and allows reporters to use proven scientific data to support their conclusions.

The Internet has opened a new route for science communication and is often your best source for up-to-date science news. In Appendix A, you will find a list of websites on which you can find reliable information about science. The warnings to be skeptical about media reports based on questionable sources applies to the Internet as well, and possibly even more so. Before you assume that a scientific "fact" that you find on the Internet is true, be sure that you know the source of the information and the possible agenda or biases of the person posting it.

How Statistics Come into Play

Englishman Leonard Courtney said in 1895 that there are three kinds of lies: "lies, damned lies, and statistics." While statistics might make the normal person's eyes glaze over, and are perhaps at times misused, they are nevertheless a science, a mathematical one. Statistics are necessary in the collections and analysis of data, as well as in the presentation of findings. In fact, statistics as a subject is applicable across all sciences. Statistics can be used to "model" data so that random variations and uncertainties are taken into account, such as in the study of demographics derived from a census of an area. Mathematical statistics can be used, for example, to determine the likelihood of something like asteroid impact on a flight to Mars. So while statistics might have been unreliable to the point of ridicule in 1895, in the current age it is a valid science that is used continuously by scientists.

> **Ss** **Science Says**
>
> "Science is simply common sense at its best; that is, rigidly accurate in observation, and merciless to fallacy in logic."
> —Thomas Huxley (1825–1895)

Science Is Always Changing

The most in-the-news scientific subject of the twenty-first century has been the idea of global warming. Putting the debate center stage before the public won former U.S. Vice President Al Gore a Nobel Prize and an Academy Award.

While global warming has been hotly debated by many prominent scientists, including the founder of The Weather Channel, a study entitled "How Well Do Coupled Models Simulate Today's Climate?" published in April 2008 in the *Bulletin of the American Meteorological Society* has provided an analysis that has convinced even some of the most prominent skeptics that global warming is real and can be attributed to human activity. Thomas Reichler and Junsu Kim from the Department of Meteorology at the University of Utah compared 50 different national and international models developed at major climate research centers around the world over 20 years, including the 2007 report of the Intergovernmental Panel on Climate Change (IPCC). In summarizing their findings, Reichler stated: "We can now place a much higher level of confidence in model-based projections of climate change than in the past."

Meanwhile, in April 2008 the head of the World Meteorological Organization announced that the recurring Pacific Ocean–based *La Niña* weather phenomenon would cool ocean waters enough to trigger a small drop in global temperatures for the year.

De **Definition**

La Niña ("the little girl") and El Niño ("the little boy") are ocean-atmosphere phenomena affecting sea surface temperature in the Pacific Ocean and causing fluctuating weather conditions around the world. La Niña results in colder temperatures and drier conditions, while El Niño generates the opposite. La Niña is usually preceded by a strong El Niño.

Scientists must often balance apparently contradictory information to build theories that explain all the observations. Research in global warming is one example of our changing and evolving understanding of nature. In the course of a single generation, global warming due to the increase in greenhouse gases has changed from a possible explanation of an observed warming to a theory that is almost universally accepted among climate scientists. Because it is an extremely complex problem, our knowledge will continue to grow and change as more data is collected.

Sorry, but That Is Not Science

"Pseudoscience" is a term used to describe knowledge that is claimed to be scientific or made to appear so, but was not developed in line with the scientific method. Pseudoscience is often used in claims by hucksters of products promising great medical benefits. If it sounds good enough, they reason, the public might believe it and buy. Television ads abound for products making fantastic claims for effortless weight loss, muscle-building without work, and "male enhancement," among others. These ads often allude to medical research, but a bit of investigation usually finds no supporting evidence for the claims.

Uncommon Sense

Just because an idea is widely accepted, that doesn't mean it is "scientifically proven." In April 2008, Dr. Dan Negoianu and Dr. Stanley Goldfarb from the University of Pennsylvania in Philadelphia made headlines around the world when they revealed that they had reviewed every published clinical study about the body's need for water. Looking for some scientific evidence of the "common knowledge" advice to drink eight glasses of water a day, they could find none whatsoever—it was a medical myth!

Other examples of pseudoscience include astrology, creation science, and claims of medical benefits of magnets or "structure-altered water." How do you know if something isn't science? True scientific research follows the scientific method and results are reported in such a way that others can repeat the experiment. If this is not the case, the results are not science.

A Look at Theories

"True science is never speculative; it employs hypotheses as suggesting points for inquiry, but it never adopts the hypotheses as though they were demonstrated propositions."

—Cleveland Abbe (1838–1916)

In science, the word "theory" has a specific meaning—a possible explanation for a phenomenon that is supported by experimental evidence. Scientists speak of the atomic theory, the theory of evolution, and the gravitational theory. They are referring, in each case, to a generally accepted description of atomic structure, or the process of changing life over time, or the functioning of the force that causes us to stay on the planet. All of these theories are based on a body of observations and experimental results.

It is an unfortunate aspect of our language that a single word can have several meanings or shades of meaning. When people say the word "theory" outside the scientific context, they may mean it as a conjecture. People refer to something "in theory" meaning in an abstract case. As a result, when a theory such as evolution or global warming is mentioned, many people think of the theories as guesses or speculation. In reality, the concepts of evolution and global warming are theories in the same sense as the concepts of the atom and gravity.

What is a scientific theory?

Scientific theories generally begin small—someone makes an observation that raises a question. For example, the theory of plate tectonics—the mechanism that moves the continents—began with an observation commonly made by school children: South America and Africa look like they could fit together like pieces of a jigsaw puzzle. Alfred Wegener asked the question: Could the two continents have once been joined? His hypothesis—that the continents float above the inner Earth—was his starting point. We now know that the continents move and the current theory of plate tectonics provides an explanation of how that happens.

A good test of a theory is its ability to predict something that has not yet been observed. For example, Albert Einstein predicted that light would be bent by a large gravitational field such as that of the sun. Specifically, Einstein's theory of relativity predicted that stars would appear to be slightly out of place in a photograph taken during a solar eclipse, due to starlight being deflected by gravity as it passed the sun. This predicted effect was observed by British astrophysicist Arthur Eddington during a solar eclipse in 1919.

Ff | **Fast Facts**

Sometimes a theory leads to predictions that are hard to accept without additional evidence. The Chandrasekhar limit, named after Indian scientist Subrahmanyan Chandrasekhar, describes the maximum mass that can be supported against gravity before matter is squeezed together enough to collapse into a black hole. Although the existence of the Chandrasekhar limit—and black holes—was predicted in the 1930s, it was not widely accepted by the scientific community, largely due to opposition by Arthur Eddington, who had great influence among his peers. Eddington agreed that accepted theories would allow the existence of black holes but did not believe they could actually develop.

A common misconception is that theories are eventually proven and then become facts. In reality, a scientific theory is never proven. A theory is a model that explains observed facts. As new data becomes available, the theory may be modified to account for the new information, or even replaced by a different theory. For example, John Dalton's theory of the atom, proposed in the eighteenth century, is very different from the current atomic theory. Our understanding of the nature of the atom has evolved as we learned about subatomic particles and the interaction of matter and energy. It is very likely that the atomic theory of the twenty-second century will be unlike the theory we use today.

A Few Theories That May Sound Familiar

Theories are an integral part of the scientific method. Some current theories, such as the atomic theory that describes the composition of matter on the small scale, have developed over a period of several centuries. Others—for example, string theory—are built on observations made within the past few decades. Other, as yet unnamed theories, may have come together today as a result of recent research.

No one can list all the theories that are used by scientists. There are uncountable theories to explain the results of the experiments performed by millions of scientists. They are constantly developed, modified, and even replaced as additional data becomes available. The atomic theory, the theory of gravity, the cell theory (all living things exist as cells), and kinetic theory (describing the motion of atoms and molecules) are a few of the basic building blocks of our current scientific knowledge.

The following theories have names that may seem familiar. These are theories that you may have seen mentioned in recent discussions about science. That does not necessarily mean that they are more important than other theories.

Evolution (Life Is Change)

In the 1760s, English civil engineer William Smith, while building canals in southern England, observed that fossils found in certain levels of rock strata were also found in different locations. Sir Charles Lyell, whose failing eyesight caused him to turn from law to geology, later made similar discoveries and came to the conclusion that Earth was millions of years old. He published his findings in the three-volume *Principles of Geology* (1830 to 1833). Charles Darwin studied Lyell's work during the voyage of the HMS *Beagle*, and was particularly impressed by his description of rock formations. These observations and discussions provided a background for the conceptual breakthrough that was to become the theory of evolution.

Darwin was also influenced by Thomas Malthus, specifically Malthus's book *An Essay on the Principles of Population*. Malthus held that nature produced a great abundance of offspring, yet over time the number of plants and animals remained roughly the same. Darwin reasoned from this that the living things better at getting food and avoiding predators were those most likely to survive and reproduce. He called this "survival of the fittest." When Darwin finally published his book after 20 years of study, he put forth the idea that all life on Earth had common ancestry and that various adaptive environmental reasons were why species changed.

Us **Uncommon Sense**

It is commonly believed that the phrase "survival of the fittest" comes from Darwin's *Origin of Species by Means of Natural Selection,* published in 1859. It was actually first used by Herbert Spencer in his *Principles of Biology* in 1864. Darwin did mention the phrase in later editions of his work. The term, as used by these biologists, referred only to the natural selection process in evolution of species. Later usages, in which "survival of the fittest" was applied to human societies and cultures, is completely unrelated to the ideas expressed by Spencer and Darwin.

The key element of the theory of evolution is that life changes as a result of natural selection. Within a species, there are always variations among individual organisms. Organisms that are best adapted to their environment are more likely to survive and reproduce. In each succeeding generation, more individuals are likely to have the traits that are most suited to the environment.

The theory of evolution, as it exists today, is built on several key elements:

♦ There are variations in form and behavior among the members of a species. Some of these variations are hereditary.

♦ Every species produces more offspring than the environment can support.

♦ Some individuals within the population have characteristics that make them better adapted to the environment than other individuals.

♦ Better adapted individuals have a greater likelihood of survival and reproduction.

♦ The favorable traits occur with greater frequency in the next generation.

♦ This process creates a natural selection for organisms possessing traits that are better adapted to the environment.

- Genetic change creates new traits that are subject to the natural selection process.

- The process of natural selection favors development of new species that are adapted to survival within an environment.

Ff **Fast Facts**

The scientific term for a "missing link" is *transitional fossil*. These fossilized remains, when discovered, illustrate an evolutionary transition heretofore not found.

Theories are not static. The current theory of evolution is very different from Darwin's original theory because a vast body of evidence has been collected since his work was published. The theory of evolution has developed and changed over time based on evidence. Many of the mechanisms of change are not yet known in detail and, within the theory of evolution, there are different theories about some of these mechanisms. Critics of evolution have latched onto some of these differences to claim that biologists cannot even agree on the existence of evolution. However, the key elements of evolution are not in dispute among scientists. It is clear that life evolves through the process of natural selection.

Global Climate Change (Not Just Warming)

It is impossible for anyone who reads or watches the news regularly to be unfamiliar with the term "global warming." This concept refers to a rise in the average temperature of Earth's atmosphere in recent—and if projections are correct, future—decades. Climate scientists, however, generally prefer the more accurate term "global climate change." This is because the effects of global warming do not present themselves as simply a steady increase in temperature every place in the world.

As atmospheric temperatures increase, many processes including global wind patterns and ocean currents can be changed. These alterations in large systems can affect the weather differently around the world. While most parts of Earth's surface will experience increased average temperatures, some regions could actually experience a decrease. Other predicted aspects of the change in global climate include alterations in patterns of precipitation. Some dry places could begin to experience devastating floods, while other regions could be subject to equally devastating droughts.

Us **Uncommon Sense**

It's not just survival that is threatened by global climate change. In April 2008, Jim Salinger, a climate scientist at New Zealand's National Institute of Water and Atmospheric Research, told the Institute of Brewing and Distilling convention in Wellington, New Zealand, that climate change portended a decline in the production of malting barley, a key ingredient in making beer. That, coupled with water shortages in certain parts of the world, could mean "pubs without beer or the cost of beer will go up." No reports were forthcoming about public reaction.

According to the theory accepted by an overwhelming majority of climate scientists, the main cause of global warming and the accompanying climate change is a change in the greenhouse effect. The greenhouse effect is an increase in Earth's temperature due to certain gases in the atmosphere, including water vapor, carbon dioxide, nitrous oxide, and methane. These gases trap energy from the sun. These gases are referred to as greenhouse gases.

What is often lost in reporting about global warming is that the greenhouse effect is a natural part of the functioning systems of Earth. In fact, it is essential to life on Earth. Solar energy passes through the gases of the atmosphere and are absorbed by soil and water. Much of the absorbed energy is radiated back into the atmosphere as long-wave infrared radiation. Without greenhouse gases, heat would escape back into space and Earth's average temperature would be about 60°F colder. The concern of atmospheric scientists, as related in the film *An Inconvenient Truth*, is that too much carbon dioxide is being released into the atmosphere by human activities. As industrializing countries, including China and India, increase their patterns of energy consumption, carbon dioxide emissions will continue to accelerate.

Very few scientists dispute the increase in the average temperature of the planet or that climate changes are occurring. The question is, is global climate change mostly caused by human activity? The United States Environmental Protection Agency (EPA) summarizes what is known about climate change on the webpage www.epa.gov/climatechange/science/stateofknowledge.html:

◆ Human activities are changing the composition of Earth's atmosphere. Increasing levels of greenhouse gases like carbon dioxide (CO_2) in the atmosphere since preindustrial times are well-documented and understood.

◆ The atmospheric buildup of CO_2 and other greenhouse gases is largely the result of human activities such as the burning of fossil fuels.

◆ An "unequivocal" warming trend of about 1.0 to 1.7°F occurred from 1906–2005. Warming occurred in both the Northern and Southern Hemispheres, and over the oceans.

◆ The major greenhouse gases emitted by human activities remain in the atmosphere for periods ranging from decades to centuries. It is therefore virtually certain that atmospheric concentrations of greenhouse gases will continue to rise over the next few decades.

Ff **Fast Facts**

Earth's climate has changed throughout history, switching between ice ages, during which much of the land surface was covered by glaciers, and interglacial periods when ice retreated to the poles, or completely disappeared. A significant change, in either direction, would make the planet less hospitable to humans and the organisms on which they depend for survival. A primary goal of climate scientists is to determine whether human activity has created a risk of such a change and develop techniques to prevent it. The fact that such changes can occur naturally does not make them any more desirable.

Evidence of heating and cooling cycles in Earth's history show that climates have changed drastically many times in the past, often very rapidly. Most atmospheric scientists are concerned now because of the large increase in carbon dioxide content of the atmosphere since 1750. There may be a "tipping point" at which greenhouse heating causes changes that increase the amount of carbon dioxide produced by natural processes. This could create a loop in which increased temperature leads to more greenhouse gas which causes more increase in temperature, and so on.

Unfortunately, understanding the effects of changes in carbon dioxide concentration is a very tough problem. Climates around the planet result from extremely complex interactions of the atmosphere and the oceans. Some of the world's largest and fastest computers are devoted exclusively to the study of climate models. The theory of global climate change is one of the most extensively studied scientific topics of recent years. The EPA website mentioned previously also lists the challenges facing climate researchers.

- Improving understanding of natural climatic variations, changes in the sun's energy, land-use changes, the warming or cooling effects of pollutant aerosols, and the impacts of changing humidity and cloud cover

- Determining the relative contribution to climate change of human activities and natural causes

- Projecting future greenhouse emissions and how the climate system will respond within a narrow range

- Improving understanding of the potential for rapid or abrupt climate change

Plate Tectonics

The theory of plate tectonics began with a simple idea and it has grown to become the backbone of modern geological science. The idea at the core of plate tectonics was introduced by Alfred Wegener in the early twentieth century in his theory of continental drift: the continents are in constant motion across the surface of Earth. Key data for the theory included the shapes of the continents, which appear to fit together like puzzle pieces; discovery of identical fossils on several continents; and mountain chains on separate continents that appear to be parts of a single chain.

Wegener hypothesized that the continents were once joined in a single supercontinent, which he called Pangaea. He proposed that Pangaea had broken apart and the continents had moved across the sea floor into the current locations. Acceptance of this hypothesis by geologists was slow because of one missing detail: Wegener could not describe a force capable of driving a continent across the sea floor.

In the 1960s, a key part of the process was found. Scientists studying the floor of the Atlantic Ocean discovered a phenomenon called "magnetic striping." As molten rock cools and hardens, minerals that contain iron tend to line up with Earth's magnetic field. Scientists discovered that strips of rock parallel to the crests of ridges in the center of the ocean alternate in magnetic polarity. This suggested that, over long periods of time, the direction of Earth's magnetic field has reversed many times. The observation also suggested that the sea floor is spreading away from the ridges.

Based, in part, on this observation, a mechanism was proposed for the movement of continents. According to the theory of plate tectonics, Earth's crust (making up both the continents and the sea floor) is actually floating on top of fluid rock. Where pieces of the crust, called tectonic plates, collide, material is pushed down into the layer beneath the crust. Where two plates move apart, molten material moves toward the surface, cools, and hardens, forming a new layer of crust.

Plate tectonics explains why some places are particularly likely to experience geological upheavals such as earthquakes and volcanoes. These events occur at the boundaries between plates, where huge masses of rock collide or grind past one another. Direct collisions between plates can cause the formation of mountain ranges as rock bends and folds. The Himalayan mountain chain, for example, has been built by the slow but continual collision of the Indian subcontinent with the plate on which the continent of Asia rests.

Ff **Fast Facts**

To understand major earthquake zones like the Pacific Coast of the United States, have a look at a map of the tectonic plates of our planet courtesy of the United States Geological Survey's map at http://pubs.usgs.gov/gip/volc/fig37.html.

Ss **Science Says**

"I am convinced that, at its best, science is simple—that the simplest arrangement of facts that sets forth the truth best deserves the title of science. So the geology I plead for is that which states facts in plain words—in language understood by the many rather than by the few."
—George Otis Smith (1871–1944)

Relativity: Easier to Understand Than You Think

One of the most famous scientific theories is Einstein's Theory of Relativity. Because the theory is based on complex mathematical calculations, rather than direct observation of familiar phenomena, the theory's reputation is one of being completely incomprehensible. It really is not that hard to understand the concepts, if you accept that the mathematics support the conclusions.

Ff **Fast Facts**

The uncertainty principle was conceived by Werner Heisenberg in 1927. It holds that the movement of a subatomic particle such as an electron can never be accurately measured, because both speed and position cannot be simultaneously assessed. "The more precisely the position is determined," Heisenberg said, "the less precisely the momentum is known in this instant, and vice versa." This principle frustrated Einstein's attempts to find a unified field theory.

Einstein actually developed two theories of relativity that combine to form the overall theory: special relativity and general relativity. The special theory of relativity, introduced in the 1905 paper "The Electrodynamics of Moving Bodies" states that all motion is relative and there is no absolute state of rest. A key concept introduced in this theory is that the speed of light is constant for all observers no matter how they are moving relative to one another. This means that time and space are perceived differently by observers depending on their motion. One of the consequences of the special theory of relativity is the equivalence of mass and energy. This equivalence, expressed in one of the most famous equations ever written—$E=mc^2$—provides the basis of nuclear weapons and nuclear power plants. Another consequence of the theory is that nothing can move faster than the speed of light.

The special theory of relativity applies to objects in motion and in the absence of a gravitational field. In the general relativity theory that came over a decade later, Einstein incorporated gravity. In this theory, the three dimensions of space and the dimension of time are combined to form a four-dimensional space-time. As objects accelerate to near the speed of light, the matrix of space and time bends to maintain the speed of light at a fixed value. Gravity is the result of this bend in space and time. Massive objects are attracted to one another by this bending—an attraction that we know as gravity.

Ss **Science Says**

"Time and space and gravitation have no separate existence from matter."

—Albert Einstein (1879–1955)

Ff Fast Facts

One of the consequences of general relativity is that fast-moving matter can create waves in space-time. These gravitational waves can radiate away from a source just as light waves radiate from a glowing body. Although gravity waves have not yet been detected, physicists have built detectors to look for the waves created by dense, spinning stars. These detectors use laser beams that are miles long to try to detect waves that create a motion that is much smaller than the diameter of an atom.

String Theory

You may have heard of string theory because it tends to capture the imagination of science reporters. Physicists who have developed string theory are looking for a pattern in the universe that goes deeper than the subatomic particles that we know and can study. Because matter and energy can be related to one another by Einstein's theory of relativity, theorists propose that at a basic level, all matter and energy consists of vibrations of loops of material or energy.

Imagine that everything in the universe was part of a constantly moving and changing pictorial tapestry, and that strings make up everything in the universe. However, these strings do not vibrate in three dimensions like a violin string. They vibrate in as many as 11 dimensions. All of the strings are the same. It is the vibrations that differ among the various types of particles. In other words, while we may be able to see atoms and molecules and electrons and photons and other subatomic particles, we cannot see the strings themselves.

 De Definition

Quantum mechanics is the branch of physics developed early in the twentieth century to explain the structure of matter at an atomic level. Curiously enough, scientists learned that subatomic matter has a wavelike structure until examined, at which time it takes on particle characteristics.

String theory is a purely mathematical idea. We cannot see the strings and no test has ever been developed to determine whether they exist. Using the equations of string theory, scientists have attempted to predict energy interactions, even the gravitational activities predicted by Einstein, and weave everything into a unified approach to *quantum mechanics*.

String theory is not universally accepted by physicists. Lee Smolin, founder of the Perimeter Institute for Theoretical Physics in Waterloo, Ontario, and Columbia University mathematician Peter Woit both published books in 2006 that stated string theory might not only be unproven, but completely wrong. Until the theorists are able to use string theory to make a prediction that can be confirmed experimentally, it is likely that many scientists will consider it to be an untested thought exercise rather than a supported theory.

Chaos Theory (That Butterfly in Brazil)

Chaos theory has captured many people's imaginations, partly because of an interesting analogy used to explain it. The movie *The Butterfly Effect* (2004) popularized chaos theory, on which it was loosely based. Chaos theory was first investigated in 1960 when meteorologist Edward Lorenz began using a computer to try to predict the weather. The term "butterfly effect" came from a paper entitled "Predictability: Does the Flap of a Butterfly's Wings in Brazil Set Off a Tornado in Texas?" The paper was presented by Lorenz in 1972 to the American Association for the Advancement of Science in Washington, D.C. The idea was simple; by flapping its wings, the butterfly made a tiny change in the system in which it existed (the jungle), which might affect, for example, the turn of a leaf, which would affect something else, and so on, eventually changing the trajectory of a larger weather system on the planet.

Although the concept of such a tiny change affecting large-scale phenomena may seem absurd, there are patterns to be found in so-called chaos. Many natural phenomena have patterns that may not be as random as they seem.

Lorenz was not the first person to study chaos. French mathematician Jacques Hadamard published a study in 1898 about the chaotic motion of a free particle gliding unaffected by friction on a surface of constant negative curvature. Chaos theory has been pursued over the years by mathematicians, who use complex formulas to try to make order out of seemingly random events and motion.

A system is classified as chaotic when: (a) it is sensitive to initial conditions, for example, the butterfly; (b) it interacts with the space around it; and (c) related points in an entire system are connected in a complex way. A small motion in the atmosphere meets these criteria. The theory can also be applied to other types of interactions, ranging as widely as the flow of fluids and the movement of money in large economic systems. Chaos theory has also been applied to studying a very large-scale phenomenon—the reasons for the particular locations of planets in the solar system.

Part 2

Physical Sciences

This part asks questions that get to the core of how the world works. Physicists look at the why and how behind force, motion, and energy. Light and sound carry information from one place to another as transfers of energy. Water and air flow from one place to another, responding to forces that we cannot see but which are capable of turning our world upside down.

Chemists look at matter. When you put two things together, they interact with one another in some way. That interaction occurs at a level too small to observe directly—the atoms that are the building blocks of everything. There are only about a hundred different kinds of atoms. How they are put together and how they affect one another form the basis for the chemist's questions.

Chapter 3

Physics—Energy and Motion

"A scientist in his laboratory is not a mere technician: he is also a child confronting natural phenomena that impress him as though they were fairy tales."

—Marie Curie (1867–1934)

We use motion and energy every moment of every day without thinking about the details. For example, throwing a baseball is a complicated process—the motion of the ball is affected by its size and shape, by gravity, by the temperature and movement of the air. The pitcher may not realize how much physics is involved in the almost instantaneous calculations needed to throw the ball to a precise spot 60 feet away.

Physics is a science that uses observation to find the rules that govern why things work as they do. It's possible to describe these rules in the complicated equations that make many people think of physics as incomprehensibly complex. It isn't necessary to use these equations, however, to understand the basic ideas behind how things move and what powers their motion.

How do spacecraft reach the moon if they use most of their fuel leaving Earth?

During the late 1960s and early 1970s, NASA's Apollo program carried astronauts to the moon 10 times (although only 6 of the missions actually landed on its surface). Each of these missions was carried into space by a Saturn V rocket, the biggest engine ever built. About two minutes into the flight, the first stage of the rocket, the largest part, dropped away after burning all its fuel. Six minutes later, the second stage of the rocket was done, having burned most of the rest of the fuel. How could the astronauts continue on a 200,000-mile journey to the moon, having burned almost all their fuel in the first hundred miles or so?

De **Definition**

Inertia is the tendency of a moving object to continue moving in the same direction at the same speed until a force acts on it, and of a stationary object to remain stationary until a force acts on it.

The astronauts relied on *inertia*, the tendency of a moving object to keep moving. Sir Isaac Newton first described inertia in his Laws of Motion. If an object is not subjected to any force, it will continue to move in the same direction forever. Inertia explains why a car will keep rolling on a flat parking lot, even if you place it in neutral gear so the engine is not affecting the wheels. Eventually the car stops rolling because of friction with the ground and the air. A spacecraft that is above the atmosphere, however, does not experience the force of friction.

A rocket does experience a force as it moves away from Earth—gravity. This is the same force that keeps a baseball thrown upward from traveling forever in a line directly away from your hand. The Saturn rocket accelerated the Apollo craft to a speed of about 7 miles per second, using a lot of energy in the process. At this speed, the forward motion of the craft due to inertia was greater than the change due to the force exerted by Earth's gravity. It continued to move away from Earth even though there was no longer any force pushing it away.

At some point in its travel, the force of gravity that pulled the Apollo craft toward the moon exceeded the force pulling it toward Earth. Then it began to actually accelerate toward the moon. On the return trip, the rocket used its remaining fuel to accelerate the craft against the force of the moon's gravity. Fortunately, because the moon is much smaller than Earth, the amount of fuel needed was much smaller.

Spacecraft can use gravity to help them move toward their destination in other ways. When the Cassini-Huygens spacecraft was launched on a course to Saturn in 1998, it did not travel directly to its destination. Instead, it was launched around the sun and close to the planet Venus. As it approached the planet, it gained speed due to the planet's gravitational pull. The craft then passed close to Earth, picking up more speed. Finally, in the year 2000, Cassini-Huygens made a pass near the giant planet, Jupiter. As it approached Jupiter, the 6-ton spacecraft's speed increased by about 36,000 mph and its course bent to send it to Saturn. The energy used to increase the speed was balanced by an equal energy change that affected the speed of Jupiter's orbit around the sun. The difference in mass means that the change in speed of the planet was quite a bit smaller than the change in speed of the spacecraft. As Cassini-Huygens sped up, Jupiter slowed by a fraction of an inch every billion years.

Ff  **Fast Facts**

The Saturn V rocket is the biggest engine ever built. During the first 8 minutes of an Apollo launch, the Saturn V generated enough power to supply New York City with all its electricity needs for 75 minutes.

How do airbags protect automobile passengers?

Automobile accidents illustrate the concept of inertia, the tendency of a moving object to keep moving in the same speed and direction and a stationary object to remain stationary. As you drive down the road at 65 mph, you move at the same speed as your car. What happens if the car stops moving suddenly—for example, by striking a bridge abutment?

A moving object also has momentum, which is the product of its mass and its velocity. During a collision, momentum is conserved, which means that the total momentum of the objects before the collision is the same as the total momentum of the objects after a collision. The inertia of the car would tend to keep it moving, but the concrete abutment has much more mass than the car. Its inertia tends to keep it from moving. Because of the high mass of the abutment, its velocity does not change much but the velocity of the car changes drastically.

Any objects inside the car, including passengers, have their own inertia and momentum. They continue moving at the same speed the car was traveling unless something interferes with their motion. Left on their own, they travel at full speed until the force of a collision with the windshield, or some other object, reduces their velocity. This is not a healthy outcome.

So how do you protect yourself? There are two ways to reduce the effects of the force of the impact—spread force over a larger area or spread it over a longer time. Seat belts spread the force over a larger area of the body. In addition, they transfer the force to the torso, which handles impacts better than the head.

Us **Uncommon Sense**

When you slam on the brakes, you seem to be thrown forward. It is easy to think that there is a force pushing you. However, the force is acting on the car around you. The passengers keep moving forward inside the car because no force has yet acted to stop their forward motion. The seat belt and airbag provide force so that the windshield does not have to.

Air bags provide even better protection. In addition to spreading the force of the collision over a much larger area—your entire upper body—air bags increase the time over which the force is applied. Although the airbag deflates in a fraction of a second after it fills, that time is substantially longer than the contact time of a person against a steering wheel or dashboard. A longer impact time means the force is not as great.

Can you really move Earth with a long enough lever?

The ancient Greek philosopher Archimedes is said to have stated, "Give me a lever and a place to stand, and I will move Earth." Was this an idle boast, or did Archimedes really see the potential of the lever? Well, perhaps it was a bit of both.

The lever is a simple machine. You exert a force at one point and it delivers a force at a different point. In the process, you change the amount of force, the direction of force, or both. However, if you increase the amount of force exerted, there is a cost. That cost is distance. Think about using a crowbar to lift a heavy rock. What happens when you push down on the bar? One end moves, maybe a foot or so. The other end lifts the rock maybe a couple of inches.

The key to operating the lever is work. In physics, the definition of work is the product of the force applied and the distance moved. The work put into the lever is the same as the work delivered at the other end. So when you decide to move the rock with the crowbar, you trade force for distance. If you have ever suffered a bad back strain by trying to exert more force than your body was designed to deliver, you know that it is a fair trade. In order to use even less force, you substitute a longer lever. Replace the crowbar with a 6-foot pry bar and you can raise the rock with a smaller push. Or you can lift a bigger rock. That's where Archimedes's boast comes in. Could he move Earth with a lever?

Ff **Fast Facts**

In physics, a machine is something that changes the direction or magnitude, or both, of an applied force. All machines are built using combinations of the six basic machines: lever, screw, inclined plane, wedge, wheel and axle, and pulley.

In principle, Archimedes is right—assuming that you could find a sufficiently long lever, and something to balance it on, and a place to stand, and you ignore the effects of gravity and Earth's motion, and so on. You get the picture. With all those caveats, what kind of lever would move Earth? Well, Earth is pretty heavy by human standards—about 6,000,000,000,000,000,000,000,000 tons. Let's give Archimedes a bit of help and provide him with a team of elephants capable of lifting 6 tons of mass. How long would the lever need to be to lift Earth by 1 inch? Because the total amount of work done by the elephants is the same as the work done on the planet, we can calculate the long end of the lever, if the short end is 1 inch. The tip of the lever would be about 500,000 times as far from the solar system as is the nearest star. There is another complication. If the elephants were to push at the speed of light (the fastest possible speed), it would take them more than 2 million years to move Earth 1 inch. So, in practical terms, it is fair to say that Archimedes could not move Earth even with his amazing lever.

Ss **Science Says**

"One of the elementary rules of nature is that, in the absence of law prohibiting an event or phenomenon, it is bound to occur with some degree of probability. To put it simply and crudely: Anything that can happen, does happen."
—Paul Dirac (1902–1984)

Why does an extension bar make a wrench work better?

When I worked in a very small chemical manufacturing plant, we used reaction vessels that held 1,000 gallons or more of chemicals. We poured solids in through a manhole-size opening on top and closed and clamped the cover. The clamps were bolted into place using a 1½-inch box wrench. Unfortunately, one of our workers was a strapping young man who could tighten the bolts around the cover so securely that none of the other employees could remove them later. The only way to open the vessel was to put a 3-foot piece of iron pipe over the wrench to make it longer. Why would the extension bar make it possible to loosen a bolt that was otherwise too tight?

> **De** **Definition**
>
> **Torque** is a force that causes a change in rotation. The magnitude of a torque is calculated by multiplying the force by the distance from the center of rotation to the point where the force is applied.

The extension allowed the rest of us to increase the *torque* that we applied to the bolt. A wrench acts like a rotating lever. As in any lever, the work done is the product of the force and the distance. As the distance from the pivot point increases, the effect of the force that you apply to the wrench also increases. The magnitude of the torque is a product of force and distance.

Those of us who could not apply as much force gained an advantage by using this rotating lever. Keep in mind, though, that there is always a cost. We applied less force but we had to push much farther to obtain the same effect.

The wrench works best when your push is perpendicular to its handle. This is because the force that adds to torque is only that part that is perpendicular to the handle. This makes sense if you think of pedaling a bicycle. The pedal arm rotates around an axle in the same way that a wrench rotates around a bolt. You feel the greatest effect of your pedaling when the arm is parallel to the ground, that is, perpendicular to your leg. When the pedal is at its lowest point, you can push down with as much force as you can generate but it will not propel the bike. As you pedal, you alternate your force from one side to the other to obtain the maximum result from the force generated by your muscles.

A wheel-and-axle combination is a kind of rotating lever. A force applied to the outside of the wheel causes a rotation. Or, a force applied to the axle can cause the wheel to rotate. In this case, the torque is very small because the distance from the pivot point is small.

Gears are an application of the wheel and axle. If force is applied to the outside of the gear, it rotates around the pivot point. Increasing the size of the gear increases the torque. The same amount of horsepower can move a large dump truck or a race car. Their engines are used in different ways, however. The truck engine generates a large amount of torque, which allows it to apply enough force to move a heavy load. The race car generates much less torque with the same applied force. However, it tends to get from one place to another in a shorter time.

Why do figure skaters spin faster when they pull their arms in?

During an ice-dancing routine, a skater starts spinning in place on the ice with her arms extended. As she draws her arms inward, raising them above her head, she spins faster and faster without exerting any force. What causes her spin rate to increase if she is not adding energy from her muscles?

When an object is moving in a straight line, it has momentum that tends to keep its motion constant. When an object is rotating around an axis, it has *angular momentum*, a tendency to keep rotating. Angular momentum does not change unless a force causes a torque that increases or decreases the momentum. For example, a toy top, once it is set in motion, continues to spin for a long time. Eventually it slows and stops because there is a force—friction— that acts to cause a torque that changes the rotation. The skater's sharp skates on polished ice minimize friction so she keeps spinning.

De **Definition**

Angular momentum is the measure of the tendency of an object rotating around an axis to keep rotating unless a torque is applied to it.

Each part of a rotating object has a moment of inertia, which is the product of its mass and the square of its distance from the axis of rotation. The angular momentum is the product of the moment of inertia and the rotational speed. The angular momentum of the spinning body is the sum of the angular momenta of each of its parts. If you are talking about an object such as a sphere, whose mass is evenly distributed around the axis, the calculation of angular momentum is fairly simple. If the spinning shape is irregular, however, such as a person, with a mass that is not evenly distributed around the axis, the calculation becomes complex. Fortunately, it is not necessary to calculate angular momentum in order to use it.

Scientists have observed that angular momentum is always conserved. That means that the total angular momentum of a spinning object does not change unless a force acts on it. Decreasing the distance from the axis of rotation decreases the moment of inertia by the square of the change in distance, so cutting the distance in half decreases the moment of inertia by a factor of four. If the moment of inertia decreases but the angular momentum stays constant, the rate of rotation must increase. As the skater brings her arms in, she spins faster and faster, going from about 2 rotations per second to 10 or more rotations per second.

Ff **Fast Facts**

A spinning object that experiences no force will continue spinning. During NASA's Gravity Probe B experiment, precision gyroscopes, spinning at 4,300 rotations per minute in a vacuum, were sent into space. During the course of the experiment, the gyroscopes turned billions of times without any additional energy being added.

The conservation of angular momentum also explains why Earth spins on its axis. According to one theory, the solar system formed from a large cloud of cosmic dust which was spinning in space. Parts of the cloud were pulled together by gravity, forming pockets of greater density which pulled even more matter inward. These clumps of matter eventually formed the sun and the planets. As the rotating material pulled closer and closer together, the total angular momentum was conserved. A large, slowly rotating cloud of matter spread out across vast areas of space became a small number of rapidly spinning balls of dense matter.

What keeps riders in their seats on a looping roller coaster?

When you get into a roller coaster that loops upside down, the attendant makes sure that your safety belt or safety bar is in place. These devices are not what keeps you in place, though. Physics takes care of that.

Acceleration is a change in direction or speed. When something moves in a circle, it is constantly changing direction, so it is constantly accelerating, even if its speed remains the same. The force that causes the acceleration toward the center of a circle is called centripetal force. Think about what happens if you swing an object attached to a string in a circle. The string remains tight and its pull forces the object to move in a circle. If you let go of the string, the object flies away. You can feel the constant force at the other end of the string.

The same thing happens when you ride a roller coaster through a loop. Because you are constantly changing direction, you are accelerating the whole time. The track is constantly pushing the car into a new direction. (It may seem strange that a stationary object can push, but in physics, forces come in pairs. The car pushes against the track and the track pushes against the car.) As a result, the seat of the car pushes you in the same direction.

When you are at the bottom of the loop, you don't notice the upward push; you are used to it because there is normally an upward push against the downward pull of gravity. As you go through the loop, your own inertia tends to keep you moving in a straight line away from the center of the circle. At the same time, centripetal force is always pushing you toward the center of the circle. As you go over the top, you feel like you should continue up into the air, but the car is pushing against you to prevent it. Even without the safety bar, you would stay in the car.

Uncommon Sense

The feeling that you are being pushed outward during a circular motion is often called "centrifugal force." In reality, there is no outward force during circular motion. What is called centrifugal force is actually a result of inertia, the tendency to keep moving in a straight line. No force is pushing outward.

Centripetal force also affects you when you go through the bottom of a dip or over the top of a hill on the roller coaster. As the roller coaster passes through a dip, centripetal force constantly pushes it upward toward the center of the curve. At the same time, the rider is pushed downward by gravity. The combination of the upward force and gravity makes you feel heavier in the dip. As you ride over top of a hill, centripetal force pulls the car beneath you toward the center of the curvature. However, the car does not push on you this time because you are above it and the force is downward. Your inertia carries you upward until the safety bar exerts a downward force. This is when you realize how important the bar is. If the force is large enough, you will feel weightless at the point where the downward force of gravity is exactly equal to your inertia upward.

How does the space station stay in orbit?

Like any other satellite, including the moon—a natural satellite—the International Space Station (ISS) orbits Earth. It follows an almost circular path around the planet almost 16 times each day. What keeps the ISS in its orbit?

The station has been in place since its assembly was begun in 1998 and has been inhabited since 2000. The space station is located in a low Earth orbit, about 210 miles above the surface. As it is built, each piece is moved into the orbit by a rocket-powered craft, such as an American space shuttle or a Russian Soyuz module. Once placed in orbit, the assembled parts continue in a course around Earth with no additional propulsion. In effect, the space station is always falling but never hitting the ground.

Remember, due to inertia, a moving object continues moving in a straight line unless a force acts on it. There is a force that acts on every object near Earth—the force of *gravity*. If you throw a ball, it does not continue moving indefinitely in a straight line because gravity causes it to move toward the center of the Earth. In a few seconds, the ball collides with Earth and the force of friction stops its motion.

The space station in its orbit also has inertia that causes it to tend to move in a straight line away from Earth. However, it is also subject to the force of gravity. It is pulled toward the center of the planet. This force is at a right angle to its inertial motion so it does not change the forward motion. It does pull the space station downward, just like any other falling object. While gravity is pulling downward, the tendency to move in a straight line would carry it away from Earth. As a result, the space station is always falling but always the same distance from the surface. Gravity is the centripetal force on the satellite, pulling it toward the center of its orbit.

De	**Definition**

Gravity is a force of attraction between any two objects. The force of gravity increases proportionally to an increase in mass of the objects and decreases proportionally to the square of the distance between the objects.

If some force were to slow the forward motion of the space station, gravity would pull it to Earth. Satellite orbits must be located above the atmosphere. Otherwise, the force of friction with the air would slow the satellite, causing it to fall. If the force of gravity were suddenly to disappear, the space station would head away in a straight line.

In fact, the space station does not hold a perfect orbit. There is some atmospheric friction, even though the atmosphere is very thin at its altitude, so it tends to lose about a mile and a half of altitude every month. When it reaches a lower altitude limit, the station is pushed several miles higher by a rocket or space shuttle.

Because the space station and everything in it is falling freely, people inside it experience weightlessness. This does not mean that they are unaffected by gravity. When we stand on Earth, the weight that we feel is not the pull of gravity. Instead, it is the upward force exerted by the ground or other object beneath us. This is a reaction force that pushes in the direction opposite the gravitational pull.

Think about descending in a very fast elevator. The floor of the elevator is dropping rapidly so the sensation of weight is reduced. You feel almost as if you were able to float. This is what is happening when astronauts experience weightlessness. The space station and all its contents, including people, are falling freely all the time. Because all of the walls are moving in the same free manner, astronauts do not feel any reaction forces to gravity. In effect, they float freely with their surroundings.

Ss **Science Says**

"A raised weight can produce work, but in doing so it must necessarily sink from its height, and, when it has fallen as deep as it can fall, its gravity remains as before, but it can no longer do work."

—Hermann von Helmholtz (1821–1894)

Us **Uncommon Sense**

Because of the term "weightless," many people believe that the space station is outside Earth's gravitational field. In fact, the strength of Earth's gravity at the altitude of the space station is about 90 percent of its strength at the surface. Earth even exerts a gravitational force on the most distant galaxies, although that force would be far too small to measure.

Why can't you cool the kitchen by opening the refrigerator?

On a hot day, there is one cool spot in the kitchen—inside the refrigerator. If you let the cool air flow from the refrigerator into the room, the temperature of the room should drop, right? Unfortunately, it doesn't work that way.

The problem boils down (no pun intended) to physics. Specifically, the laws of thermodynamics say that there is no way for you to get something for nothing. So, if you make one place cooler, you have to make another place warmer. As you cool your perishable foods, you heat up the kitchen. A refrigerator is a kind of heat pump. It removes heat from the inside of a box and then it has to dump the heat somewhere else, outside the box. In order to move heat, it compresses a gas. The gas gets warm and then it is cooled by a fan. The heat goes out of the back of the fridge. Place your hand behind your refrigerator when it is running and you should feel a flow of warm air. The cooled gas is then allowed to expand inside coils on the cool side of the box. When gas expands, it gets even cooler than the air around it. Heat flows from the air to the coils. Then the gas is again compressed and cooled, starting the cycle again.

Left to its own devices, heat only flows from warm places to cool places. To make it go the other way, making the cool inside of the fridge even cooler, requires work. To do that work requires a machine and it is impossible for a machine to be 100 percent efficient. That means that as it runs, the machine dumps the heat from the inside of the refrigerator and it dumps even more heat due to its inefficiency. In fact, at its best, a refrigerator has an efficiency of about 40 percent. The additional heat comes from running the motor and compressor.

So what does this all mean for kitchen cooling? Initially, you will dump cool air from the refrigerator into the kitchen, dropping the average temperature of the room by some amount. Immediately, however, the thermostat in the refrigerator detects that the temperature inside has risen. The motor kicks on, the compressor kicks on, the system starts cooling the air in the box but it is quickly warmed by air coming through the open door. The final result, thanks to the inefficiency of the heat pump and the dumping of the same heat that has been removed, is that the room heats by about 2½ degrees for every degree that it is cooled.

a gas is not the only process that can be used in
that involves an energy change could possibly be
studying magnetic refrigeration based on changes
hen it is in a magnetic field compared to its energy

etween AC and DC electric

moving electrons, tiny particles with an electric
ns carries energy that can be used to do work.
the energy can be tapped. When you turn on a
bulb resists the flow of electrons. This resistance
ms in the filament and energy is produced as heat
nts work in heaters and stovetops the same way.
verted to mechanical energy in an electric motor
current and a magnetic field to produce motion.
l devices labeled for AC (alternating current) or DC
ference between the two types of current?

from one place to another in a single direction. For
ashlight, electrons flow from the negative pole of
ulb, and then to the positive terminal. As they pass
lose energy and light is emitted. In an alternating
w from point to point. Instead, they move back and
direction in a cycle of 60 times each second. When
ight bulb, the electrons move back and forth within the
element, losing energy due to resistance to their motion.

Early in the history of electric power, both types of current were used to transmit
electricity. The first power plants used direct current. However, alternating current
is able to transmit energy over much greater distances without losing energy. In addition, AC voltage can be changed very easily using a transformer. It is more efficient

to transport the power over long distances at very high voltage, but for use inside homes and businesses, a lower voltage is necessary for safety. A series of transformers in the transmission and delivery system reduces the voltage from production to use.

Ff Fast Facts

Although the current flows instantaneously through a circuit when it is turned on, each individual electron moves slowly within the circuit. In a DC circuit, such as the one between a car's battery and its headlights, each electron may move about 1 millimeter per second, pushing along the electron ahead of it. In an AC circuit, the electrons do not move through the circuit at all. Each electron moves back and forth along the wire.

For some uses, it does not matter whether the power source is AC or DC. Electric lighting, for example, which is based on resistance to current, works with either type of power. Other applications require a specific type of electric power. Electric motors are designed to work with either AC or DC current and will not operate if the wrong source is used.

Direct current is normally used in low-voltage applications. Batteries and solar power systems can only produce direct current. Most electronic devices also require DC power. In order to operate these devices on the household AC current, an adapter is needed. AC adapters convert the alternating current into a direct current, which operates the device or charges the battery without damage.

"I had an immense advantage over many others dealing with the problem inasmuch as I had no fixed ideas derived from long-established practice to control and bias my mind, and did not suffer from the general belief that whatever is, is right."

—Henry Bessemer (1813–1898)

Chapter 4

Physics—Light and Sound

"The most important fundamental laws and facts of physical science have all been discovered, and these are now so firmly established that the possibility of their ever being supplemented in consequence of new discoveries is exceedingly remote."

—*Albert A. Michelson (1852–1931)*

The science of physics includes the study of waves. You know about waves from watching the ocean or the surface of a lake or pond. Other types of waves include sound and electromagnetic radiation, such as light, radio waves, microwaves, and x-rays. An earthquake occurs when waves traveling below the surface cause the ground to move up and down or side to side.

Waves can be described mathematically, and the same equations can be used for all kinds of waves. But it is not necessary to understand the equations to observe how waves function. The key thing to know about a wave is that it transfers energy from one place to another. Electromagnetic waves can carry energy through space. Electromagnetic waves include the light that you see; radio waves that carry information to your television, cordless phone, and car radio; x-rays that make an image as they pass through your body; and the microwaves with which you cook dinner. These waves are all related and they all carry energy.

Mechanical waves include sound, water waves, and earthquakes. Mechanical waves don't travel through space; they travel through a medium such as air, water, or Earth itself. The energy of mechanical waves is carried by the motion of the particles of the medium. If you watch "the Wave" course through the stands of a large football stadium, you can observe the energy of each individual particle as the fans carry their energy around the stadium.

Why is the sky blue?

When astronauts in the International Space Station look out the window, they don't see a yellow sun in a blue sky. Instead, they see a white sun against a black background. So why does the sky look blue from where you stand on the ground? And for that matter, why do the puffy cumulus clouds floating across the blue sky appear to be white (except when the sun is rising and setting)?

Light travels in waves of different lengths. The *wavelengths* of visible light are very short—about one ten-thousandth to one one-thousandth of a millimeter. The longest waves in the familiar spectrum of visible light are red and the shortest are violet. The sun produces light across the entire span of the spectrum. When all of the wavelengths of the visible spectrum are seen at once, the light appears white. That's why the sun looks white to an observer on the space station. The sun also produces ultraviolet radiation, which has a shorter wavelength than that of violet light, and infrared radiation, which has a wavelength longer than red light. However, since you can't detect these types of radiation with your eyes, they don't concern us in this discussion.

De | **Definition**

The **wavelength** of a wave is the distance between one crest of a wave and the next crest. The length of a wave of red light may be 640 billionths of a meter while the length of an ocean wave may be many meters.

When light reaches Earth, it has to pass through the atmosphere, the layer of gases that surrounds the solid and liquid parts of the planet. The atmosphere is full of dust particles and droplets of water vapor. When light hits these particles, it reflects and changes direction. As the light passes through the atmosphere, it is reflected over and over. That's why sunlight can reach places that are not in a direct line to the sun. Water droplets are larger than the light waves so they tend to reflect every wavelength. That explains why a cloud looks white—the droplets reflect all colors of light and the combination looks white.

Most of the atmosphere consists of molecules of nitrogen and oxygen, which are much smaller than the wavelength of light. When light strikes these molecules, it does not reflect in the same way that it reflects from water droplets. However, when light strikes them, the molecules occasionally absorb the energy of the light wave. After a while (a very tiny fraction of a second, generally), the molecules emit light at the same wavelength. While all of the light from the sun is traveling in the same direction, the light emitted by a gas molecule can travel in any direction in a straight line away from the molecule. The light is scattered in all directions. This is called Rayleigh scattering, named after Lord John Rayleigh, who first described it.

So how does this explain the color of the sky? It turns out that the shorter the wavelength of the light, the more likely it is to be scattered by Rayleigh scattering. Long wavelengths, such as red and yellow, are much less likely to be scattered than short wavelengths, such as violet and blue. If there were no atmosphere the sky would look black, just as it does in space. What you see when you look at a blue sky is blue sunlight that has bounced around in the air until it is coming at you from all directions.

Why, then, doesn't the sky look violet? After all, violet light has an even shorter wavelength than blue light. There are two reasons for this. First, the sun emits more blue light than it does violet light, so the scattered blue light tends to be more visible to the eye because there is more of it. Second, your eye is much more sensitive to blue light that it is to violet light. These effects combine to make the sky look blue.

Ff | **Fast Facts**

The sky often looks red in the early morning and late evening. During these times, sunlight must travel a greater distance through the atmosphere to reach the observer. Shorter wavelengths are scattered along the way, so the light that reaches your eye is the long-wavelength red light.

Why does a spoon in a glass of water seem to bend?

Place a spoon in a clear glass that is half full of water. Look down at the spoon from different angles and you will see that its stem seems to bend or break where it enters the water. You can pull the spoon out and see that it is not bent, so what happened? The apparent bending occurs because light does not always travel at the same speed.

"Wait a minute," you say, "I always heard that light travels at a constant speed. Wasn't that the basic assumption of Einstein's theories?" Well, Einstein based his theory on the speed of light *in a vacuum*, which is constant at about 186,000 miles per second. When light passes through a transparent material, such as air, water, or glass, it does not travel at the same speed as in a vacuum. Light moves through materials at a slower speed than it moves through a vacuum.

In addition, the speed of light in the material depends on the nature of the material itself. For most purposes, the speed of light in air can be considered to be the same as the speed of light in a vacuum. In water, however, there is a large difference. Light travels through water at about three fourths the speed at which it travels through a vacuum. The speed of light in glass depends on the composition of the glass, but generally it is about two thirds the speed of light in a vacuum.

The apparent bend in the spoon's stem comes from this change in light speed as the light travels between the water and the air. You see the spoon because rays of light are reflected from its surface. When these rays of light pass from the water into the air, they travel at a different speed.

When parallel rays of light travel through the water, some reach the interface first and begin to move faster. Others take longer to reach the air, so their change in speed is delayed longer. As a result, the path that the light follows is bent. This bending of the light path causes the spoon to appear to bend. If you go fishing and see a large fish in the water, you cast your bait in front of the place where the fish appears to be swimming. Because of refraction, the fish seems farther away than it actually is.

De | Definition

The **refractive index** of a material is the ratio of the speed of light in a vacuum divided by the speed of light in the material. The refractive index determines the angle of refraction as light enters and leaves the material.

The amount of bending depends on the change in light's speed as it passes from one material to another. The angle of change when light passes from air into a transparent material is called the *refractive index* of the material.

We use the change in the angle of light whenever we wear corrective lenses. If the lens of your eye does not focus an image at the right place on the retina, the object that you are looking at appears blurry. Eyeglasses use lenses that are carefully shaped to bend light as it enters and leaves the glass to change the point where light rays come together. The shape of the lens determines whether the image moves forward or backward in the eye. The correct lens places a sharp image on the retina.

Ff **Fast Facts**

Diamond has one of the highest refractive indices of natural materials. The speed of light in a diamond is only about 40 percent of the speed of light in air. The large refraction in diamond, combined with reflections from the cut facets, give the diamond its characteristic sparkle.

Why does the pavement ahead look wet on a hot, dry day?

What comes to mind when you hear the word "mirage"? If you grew up watching cartoons, you probably picture a thirsty, haggard man crawling across the desert, looking at an oasis, complete with palm trees and a shimmering lake. Is it a figment of a mind driven by thirst or something else? Many people think of a mirage as a delusion created by stress, thirst, and exhaustion. While the mind may add an illusion to it, a mirage is an actual phenomenon of light that can be recorded by a camera.

You can often see a mirage as you drive along the highway on a hot day. Watch the road far ahead and you see puddles that disappear as you get closer. No matter how fast or how far you drive, you never reach the puddle. The mirage is caused by the same trick of light as the bent spoon—refraction. The speed of light in air changes slightly as the temperature changes.

Mirages occur when a sharp boundary forms between two layers of air with different temperatures. If the air just above the ground is much warmer than the air above

Us **Uncommon Sense**

Because of their popular portrayal, it may seem that a mirage is specifically a hot desert phenomenon. Actually, a mirage can occur under any conditions in which a temperature gradient can bend light. Mirages can even be seen above the ice in Antarctica.

it, light bends upward. The puddle that you see on the road ahead is actually light from the blue sky and clouds in the distance. Your mind, however, treats the image as if it occurred in a straight line from your eyes, not accounting for the bending of the light. It appears as though the light from the sky is reflecting from the surface of a puddle or lake. This type of mirage is usually seen on hot days when the sun heats the air above the pavement, but it can occur whenever the air near the ground is much warmer than the air above it. Sometimes you can see images of objects other than the sky in a refracted image. Try looking at the air above a dark-colored car that has been sitting in the sun on a hot day. You may see an inverted image of the storefront on the other side of the street.

There is another type of mirage that occurs when cold air lies beneath warmer air. Then light can bend downward so that a distant object appears above its actual position. When this happens, an object such as a ship on the ocean can be seen even though it is beyond the horizon. Sometimes mountains or buildings can be seen floating above the ground.

What makes a rainbow move as you drive?

If there is a pot of gold at the end of the rainbow, you would expect to have a better chance of reaching it in a car than on foot. Unfortunately, the rainbow will fade before you reach it, no matter how you travel. The physics of light guarantees that no one will ever arrive at the point where the rainbow touches the ground.

To understand how a rainbow forms, consider what happens when sunlight passes through a prism. As the light crosses the boundary between air and glass, it is refracted, or bent so that it travels at a different angle. When the light leaves the glass, it is refracted again. The prism shows the colors of the spectrum because different colors of light are bent by different amounts. Although all light travels at the same speed in a vacuum, the speed of light in matter depends on its wavelength. Because the angle at which light is refracted is a function of the change in speed, each wavelength bends by a slightly different amount than any other wavelength. The result is the familiar separation of white sunlight into its component colors.

Rain consists of many spherical drops of water (not the teardrop shape that is often drawn for raindrops). When sunlight passes from air into a drop of water, the light bends. Some rays of the light reflect from the back of the drop. When these rays pass from water to air, they bend again. Each wavelength (color) of light bends at its own characteristic angle and comes out of the drop at a different angle. A ray of red light that enters the drop can reflect and leave at an angle of 42° compared to the entry ray. A ray of blue light exits at an angle of 40°. If you are standing in the path of one of the rays that exits the drop, you will see a light of that color. If there are many drops of water, as in a rainstorm or a fine mist from a hose, the light you see from each drop depends on your position compared to that drop. You see the sum of these light rays as a rainbow.

There are a couple of implications of this explanation of rainbows. First, a rainbow will only form when the sun is behind you. Otherwise there is no light ray to refract and reflect back to your eye. Second, you can never reach the rainbow, no matter how quickly you move toward it. The rainbow is an image formed by water drops in your line of sight. As you approach the drops, you see them from a different angle and no longer detect the light rays of the rainbow. However, light from more distant drops may still reach you, so you find that the rainbow recedes as you move.

Ss | **Science Says**

"Since the rainbow is a special distribution of colors (produced in a particular way) with reference to a definite point—the eye of the observer—and as no single distribution can be the same for two separate points, it follows that two observers do not, and cannot, see the same rainbow."

—William J. Humphreys (1862–1949)

Ff | **Fast Facts**

What happens if light does not leave the drop after being reflected? It can reflect from the front of the drop to the back and reflect again toward the front. When this happens, the ray leaves the drop at a greater angle and you see a double rainbow. The outer rainbow is not as bright and its colors are reversed.

Why do colored objects look different under different lighting?

Have you ever looked at a paint sample in the store and then taken it home only to find that it looks completely different? The color of the sample shouldn't change during the drive home, so what happened?

How you perceive the color of an object depends on two different factors: the wavelengths of light that are coming from the object to your eyes, and how your eyes and brain interpret those wavelengths. Some objects—for example a light bulb, television, or electric heating element—emit light. Other objects, such as a wall, a leaf, or this book, reflect light from other sources.

Ss **Science Says**

"The main source of all technological achievements is the divine curiosity and playful drive of the tinkering and thoughtful researcher, as much as it is the creative imagination of the inventor."

—Albert Einstein (1879–1951)

When the emitted or reflected light reaches your eye, it interacts with cells in your eye. There are four types of cells. Three types of cone cells are sensitive to different ranges of light wavelengths. These cone cells used to be called blue, green, and red cones but they are actually sensitive to a wide range of wavelengths centering around these colors.

There is a great deal of overlap in the wavelengths that stimulate the cells, so a single wavelength may be detected to different degrees by two or even three types of cone cells. That's where the brain comes in. Your brain converts the combined signals from the cone cells into the perception of a particular color. It does not matter whether the signals come from a single wavelength or a combination of many wavelengths. A particular set of stimulations is perceived as a particular color. That explains why a computer monitor can produce all possible colors using a combination of three colors of dots. A fourth type of cell in the eye, the rod cell, does not distinguish color. Rods generally work only when the level of light is low. Cone cells do not work well in low light, so you tend to see in black and white when it is dark.

So why do objects seem to have different colors in different types of lighting? Picture a green leaf on a sunny day. Why does the leaf look green? When sunlight strikes the leaf, molecules inside the leaf absorb some of the energy of the light. Some wavelengths are absorbed more than others. This is the source of energy that a plant needs in order to grow. The wavelengths of sunlight that are not absorbed by the leaf are reflected. When you look at a leaf, you see the light that is reflected from its surface and you perceive this light as green. What happens if the light that strikes the leaf is not sunlight, but rather light from a red stage light that produces only light in red wavelengths? In this case, the leaf absorbs all of the light that strikes it. When no light is reflected or emitted, an object appears to be black.

There are several kinds of artificial light used in buildings. Although the light they produce generally appears to be white, none of these light sources produce a spectrum of light that exactly matches sunlight in distribution and relative intensity of wavelengths. They also differ from one another. For example, incandescent light (light bulbs) tends to produce more light in red and yellow wavelengths compared to sunlight. Standard fluorescent tubes tend to produce light that has a greater proportion of violet and blue wavelengths. The light that is reflected from an object varies depending on the wavelengths of light that strike it. That is why a paint chip may appear different under different types of lighting. Some stores that sell paint offer a place where the colors can be observed under varying types of light.

Ff | **Fast Facts**

Animals do not necessarily see things the same way that you do. Some birds and fish have four types of cone cells, so their color vision may be more precise than ours. Owls, however, like many other nocturnal animals, have no cone cells. While they have excellent vision, they perceive the world in black and white. Bug zappers use ultraviolet lights to attract mosquitoes because they see the short-wavelength violet and ultraviolet light, but not yellow or orange. And you know how a matador uses a red cape to entice a bull to charge? Well, the color is for the spectators. Bulls have only two kinds of cone cells and they do not perceive red as a color.

Why do you see lightning before you hear thunder?

During a thunderstorm, you see a flash of lightning on a nearby hilltop. Several seconds later, you hear a loud clap of thunder. Which was produced first—the lightning or the thunder, or did they occur at the same time? Most people today know that the thunder and lightning occur at the same time, but that has not always been clear. In fact, about 2,300 years ago, Aristotle thought that thunder occurred when trapped air was forced out of clouds. Lightning was thought to follow the thunder when this released air burned. So, lightning occurred after thunder and it only appeared to come first because we see it first. Later, many people found it obvious that the lightning occurred first because we see it first.

There is a common belief that lightning does not strike the same place twice. There is no scientific basis for this belief. Lightning strikes the highest points of very tall buildings, such as the Empire State Building, dozens of times every year. If you watch closely, lightning sometimes seems to flicker several times. These flickers are repeated strikes on one spot within about a second.

We now know that thunder and lightning occur at about the same time. Lightning is an incredibly hot electrical discharge that occurs when opposite charges build up inside clouds or between clouds and the ground. The air around the discharge is heated almost instantaneously to about 50,000°F (five times the temperature of the sun's surface). When the air is heated so quickly, a violent expansion creates a shock wave in the air around it, much like that of an explosion. This wave travels through the air, carrying vibrations to your ears, where they are perceived as the sound of thunder.

So why do you see the lightning first? Light travels through air at about 186,000 *miles* every second. Sound, however, is much slower. A sound wave travels through the air as vibrating molecules bump into their neighbors, transferring energy. This process takes time, so a sound wave moves about 1,000 *feet* in that same second (the speed varies a bit, depending on temperature and the amount of water vapor in the air). You can estimate the distance to the lightning strike based on this difference in the speed of light and sound. Count the seconds between the time you see the lightning and the time you hear the thunder. Divide the number of seconds by 5 to find the approximate distance, in miles, to the lightning strike.

You may have noticed that sometimes thunder sounds like a sharp clap and other times it is a dull rumble. This difference can also be explained by the time that it takes for sound to reach your ears. Because the electrical discharge is essentially instantaneous along the length of the lightning bolt, the sound wave is created instantaneously as well. If the lightning strike is very close to your position, the sound waves do not spread very far before reaching your ear. Then you hear a loud crack or bang. However, if the lightning strike is far away, sound waves from the top of the lightning bolt may travel much farther than those at the bottom. Because the sound wave may be several miles across, you hear a long rumble of thunder.

On average, there are about 6,000 flashes of lightning (and peals of thunder) every minute around the planet. Thunder and lightning always occur together but you may occasionally detect one without the other. The sound of thunder can travel about 10 miles, but the light from a lightning strike may travel much farther, especially if it reflects off the water vapor in clouds, so you see the flash but hear no thunder. In other circumstances, the flash may be hidden by dense clouds so you hear distant thunder without seeing the lightning.

What causes an echo and why can't you hear one in your living room?

If you have ever been in a wide canyon with distant walls, or even a large empty room such as a gymnasium with solid walls, you have probably heard an echo. A shout is repeated, sometimes more than once. What causes an echo and why can't you hear one in your living room?

A sound wave travels through air as a series of compressions of the gases that it travels through. What happens when these compressions strike a flat object such as a wall? If the wall is soft and absorbent, such as the tapestry or cork coverings on the walls of a concert hall, at least part of the energy of the wave is absorbed. If the wall is smooth and hard, like the rock walls of a canyon or the concrete walls of a large gym, the waves may reflect just as light waves reflect from a mirror. An echo occurs when sound waves reflect and return to your ear.

An echo occurs whenever a sound wave reflects. However, if the echo occurs less than one tenth of a second after the original sound, you generally cannot separate the two sounds, so you don't hear the echo. At normal room temperature, sound travels in air at about 750 mph (or 340 meters per second). That means you will normally only hear an echo when sound bounces off an object that is at least 17 meters away. This will be the case in a large gymnasium but not in the typical living room.

Ff Fast Facts

In the early years of the twentieth century, it was widely reported that a duck's quack does not echo. Researchers at the University of Salford, England, tested this idea by placing a duck named Daisy in a chamber designed to measure the reflections of sound. They determined that a quack, like any other sound, does indeed reflect, causing an echo. They proposed three possible explanations for the origin of the belief: 1. the quack is usually too quiet for its echo to be detected; 2. ducks don't quack near reflecting surfaces; 3. it is hard to hear the echo of a sound that fades in and fades out.

You can use the echo to determine the distance between you and a mountain or a canyon wall. Make a loud, sharp sound and measure the time until you hear the echo. Knowing that sound travels at 340 meters per second, you can find the distance in meters by multiplying the number of seconds by 340 and then dividing by 2 (remember that the sound has to travel to the wall and then the echo must return the same distance).

Echoes allow bats to fly very quickly in the dark without colliding with walls, light poles, or other objects. Bats make very high-pitched sounds, too high for humans to detect. If you look at pictures of bats, you will see that they generally have large ears, the better for hearing echoes. Using the time between the sound and the echo, a bat can accurately determine the location of an object and turn to avoid it. Fortunately for bats, they can determine the echo in times far less than 0.1 seconds.

Why does the sound of a racecar engine change as it passes?

In the stands of a stock car race, or even on a television broadcast of the race, you can hear a change in the sound of the car as it passes. As the car approaches, the powerful engine makes a high-pitched whine. Suddenly, as it passes your observation point, the engine sound drops to a low rumble. Why would the position of the car affect the sound of its engine?

De | **Definition**

Frequency is the measure of the number of waves passing a point in a unit of time. Frequency is often measured in hertz (Hz), or waves per second. The frequency of visible light ranges from 450 trillion Hz to 750 trillion Hz. Human ears can detect sound in a frequency range of 20 Hz to 20,000 Hz. Ocean waves typically occur at frequencies in the range of 0.05 Hz to 1 Hz.

The Doppler effect, named for Austrian scientist Christopher Doppler, who described it in 1842, is a change in the *frequency* of a wave due to motion of the source relative to the observer. You can detect the Doppler effect whenever an object that makes a constant sound is moving relative to your position. Examples include the sound of a whistle on a passing train, a siren on a police car, and even a mosquito flying by.

You can picture the Doppler effect by thinking of sound waves at a specific wavelength as compressions in the air that occur at a constant interval. As each compression enters your ear, it causes sensors to vibrate and send a message to your brain that is interpreted as sound. Shorter wavelengths have a greater frequency—that is, compressions occur at a greater rate. A higher frequency corresponds to a higher pitch. If the source of the sound is moving toward you, each compression starts at a point that is closer to you than the previous compression. As a result, succeeding waves reach your ear more frequently than they are emitted by the source, so the sound has a higher pitch. As the source moves away from you, the waves are spread out and reach you with less frequency. The pitch of the sound is then lower. The same thing would happen if the source of the sound were stationary and you were moving quickly toward or away from it.

The Doppler effect applies to all kinds of waves, not just sound. For example, when a police officer uses radar to measure your speed, that is an application of the Doppler effect. The radar system sends radio waves at a constant frequency toward your car. The wave that is reflected back is shorter if you are moving toward the detector. By comparing the frequency of the returning wave to the frequency of the original wave, a computer in the radar detector can calculate your speed. Other applications of radar include tracking the motion of airplanes and ships, and even monitoring a thunderstorm by measuring the motion of water droplets in a cloud. Astronomers even use the Doppler effect to measure the expansion of the universe. The light from distant galaxies is shifted toward the red end of the spectrum because the distance between them and our galaxy is increasing at rates close to the speed of light.

Ff | **Fast Facts**

The Doppler effect can be used in medicine. Doctors use ultrasound, sound waves that have a frequency higher than those detected by the human ear, to measure the speed of blood flowing inside the body. The waves reflect off the moving fluid and return to the detector. The change in frequency indicates the speed of the blood flow.

Physics—Fluids

"If there is magic on the planet, it is contained in Water."

—Loren Eiseley (1907–1977)

What happens if you push on the surface of water? It flows away from the point at which you are pushing and your finger fills the space formerly filled by water. If the water is frozen, however, this does not happen. Substances that flow under stress are called *fluids*. Generally, the term applies to liquids and gases. In solids, atoms or molecules are held tightly to the atoms and molecules around them. It is difficult to move one particle away from nearby particles. In gases and liquids, these particles are easily separated from one another. As a result, the physical properties of fluids are often very different from those of solids.

The key difference between materials that are fluid and those that are not fluid is the ability of particles to move independently. Traffic engineers use the physics of fluids to analyze the flow of vehicles on busy highways. The ebb and flow of traffic can often be explained by the mathematics of fluid flow.

> **De** **Definition**
>
> A **fluid** is any substance that continually deforms under stress, no matter how small the stress that is applied. Although all liquids and all gases are fluids, some fluids are neither liquid nor gas.

What is the shape of a raindrop?

Everyone knows the shape of a raindrop, right? It is rounded on the bottom, tapering to a point at the top. Wrong. This classic shape is familiar to everyone—it shows up in books, illustrations, the nightly TV weather forecast, and sometimes even in science textbooks. While it is true that a drop forming on a dripping faucet has approximately that shape as it clings to the spout, a free-falling drop is a different story. It is hard to look at falling rain, though, and analyze the drop as it passes your face. High-speed photography has come to the rescue.

In reality, raindrops don't have a "raindrop" shape or anything resembling it. The shape of a raindrop depends on its size and how fast it is falling, but it tends to be roughly spherical. Look up at the sky on a cloudy day and you see a collection of very small water droplets. They form when water collects on small pieces of dust. The molecules of water are attracted to one another and pull together into a shape that keeps them closest. This shape is a sphere. At the beginning, these spheres are very tiny, ranging from about one one-thousandth to one twentieth of a millimeter in diameter. As these small droplets move around, they bump into one another and form larger drops. Eventually these drops become too heavy to be supported by the air around them and they fall toward the ground.

> **Ff** **Fast Facts**
>
> The speed at which a raindrop falls is determined by the pull of gravity on the water and friction between the water and the air around it. The smallest drops hang in the air as fog, the largest can reach speeds of about 20 mph before the effects of wind resistance break them apart.

As rain falls, the drops continue to bump into one another, sometimes clinging together to grow larger and sometimes breaking apart. The size of raindrops varies from about 1 to 4 millimeters in diameter. The smallest of them tend to keep a spherical shape as they head for the ground. Larger drops are affected by the force of air pushing against them from below. Because of *surface tension* from the attraction of water molecules on the outside of the drop, the top of the drop keeps its rounded shape. The bottom flattens, however, due to the air pressure. As the drops grow even bigger, the flat bottom pushes inward and becomes a depression, sort of like a soccer ball that has lost all its pressure and then had one side pushed inward.

 **De** Definition

Surface tension is a property of a liquid that causes its surface to act as a sheet. Surface tension is a result of forces that cause particles (atoms or molecules) of the liquid to be attracted to one another.

As raindrops grow, the depression becomes bigger so they are thinner and thinner in the middle. Eventually, the drop becomes too large (greater than 4 mm in diameter) and it splits into two smaller drops.

Why does a bicycle pump get hot when you inflate a tire?

Try pumping up a tire with a bicycle pump by rapidly pushing up and down on the handle of the pump several times. If you feel the barrel of the pump, you will find that it has become quite warm. One possible explanation for the heat is friction. However, you can try the experiment in reverse. Close the end of the hose with the handle all the way down. Now pull the handle up very quickly and you will find that the barrel of the pump is cold. Friction won't explain that temperature change, so what is happening?

Gases, such as air, are different from liquids and solids in that they are compressible— that is, they can be pushed into smaller volumes of space. Gas molecules tend to fly around freely with a lot of space in between. The temperature of a gas is a result of this motion and heat is transferred as particles bump into one another. You detect the temperature of air when its particles strike your skin, transferring energy from the particles to you. On a hot day, faster-moving molecules hit your body more often, so more energy is transferred. On a cold day, the particles are lethargic by comparison, so there are fewer collisions, each transferring less energy.

When you compress air in a bicycle pump, your muscles transfer energy to the handle, which in turn transfers energy to the molecules of air in the pump. This additional energy makes the molecules move faster. As they are compressed into a smaller space, they also collide more often with the wall of the pump, so they transfer more energy to the metal wall and it becomes hot.

Ff **Fast Facts**

The diesel engines in large trucks do not need spark plugs. Inside the engine, air is compressed to about one twentieth of its original volume. This compression heats the air so much that, when fuel is injected, the mixture ignites spontaneously without a spark.

The temperature, pressure, and volume of a gas are related to one another. Temperature increases as the volume occupied by the gas decreases and/or its pressure increases. When you compress it, air gets hotter. When you expand air, it gets colder as you remove energy. The particles move slower and don't collide as often.

It is this relationship between pressure, volume, and temperature that explains the temperature difference between the top and the bottom of a mountain. Air at the bottom of the mountain is compressed by the weight of the air above it, so it tends to be warmer.

If a mass of air rises in the atmosphere, the pressure drops and the volume of the air expands so it becomes cooler. If a mass of cool air drops from the upper atmosphere to a lower level, it becomes compressed. The molecules are pushed closer together and the air becomes warmer. These changes in pressure, temperature, and volume of air masses can release large amounts of energy that fuel thunderstorms and even hurricanes.

Ff **Fast Facts**

Air is compressed by the mass of air above it, causing the air at lower elevations to have a greater density (density is the mass of material divided by its volume) and higher temperature, if all other factors are equal. The temperature at the bottom of the Grand Canyon is normally about 25°F higher than the temperature at the canyon's rim due to the difference in air pressure.

Why does an oil tanker float?

Everyone knows that some things float in water and some sink. A small stone, weighing a few ounces, sinks. Meanwhile, an oil supertanker, which weighs hundreds of thousands of tons, has no trouble staying afloat. What keeps the steel ship on top of the water but not the stone?

The answer to that question is based on the physics principle called *buoyancy*, a force that pushes upward on an object in a fluid. Because water is a fluid, an object placed on top of it pushes some of the water out of the way. The amount of force used to push the water is equal to the force of gravity on the object. The object—a stone or a tanker—pushes water out of the way until the mass of water moved is equal to the mass of the object itself. The buoyant force is equal to the mass of the water that an object displaces.

De | **Definition**

Buoyancy is an upward force exerted on an object in a fluid as a result of the difference in pressure exerted by the fluid on the top and the bottom of the object.

No matter what the weight of the object, it will float if the buoyant force is greater than the pull of gravity. That means that a stone, which has more density than water, sinks to the bottom. It continues falling through the water, pulled by gravity, until it reaches the bottom. If the object is less dense than water, it sinks until it has displaced an amount of water equal to its weight. After that, the buoyant force causes it to float.

Ships are full of air pockets and voids inside the hull that cause them to reach this displacement before the water reaches the top of the boat. Although steel is denser than water, the shape of the ship and the empty spaces inside make the ship less dense than water. However, if you replace the air inside the ship with water, it quickly sinks. Watch *The Poseidon Adventure* to see what happens as density increases.

Ss | **Science Says**

"Any solid lighter than a fluid will, if placed in the fluid, be so far immersed that the weight of the solid will be equal to the weight of the fluid displaced."
—Archimedes (287 B.C.E.–212 B.C.E.)

Materials that are less dense than water don't need to be hollowed out in order to float. Place a wooden block and a foam block in a container of water. Both materials will float. However, much more of the wooden block will be beneath the surface because it has a greater mass and displaces more water.

How does a submarine rise and submerge?

Ships float, rocks sink—and then you have submarines. They can float on top of the water or dive beneath it. How do submarines go up and down?

The buoyant force occurs because the pressure in a fluid increases with its depth. You can feel this pressure difference in your ears when you dive into a pool. The deeper you go, the more the water pushes on your eardrums. When an object is placed in water, the water pushes on it. The push is stronger as depth increases, so the overall force is upward, opposite the force of gravity. When you hold an object underwater, it feels lighter than in air because of this upward force.

Submarines use this principle when they move up or down in the water. A submarine has a double hull. The space between the two hulls forms tanks that can be filled with water or air. When these ballast tanks are filled with air, the buoyant force is greater than the force of gravity so the submarine floats on the surface. As the air in the tanks is replaced with water, the weight of the ship increases although its volume remains the same. When the downward pull of gravity exceeds the buoyant force, the submarine sinks toward the bottom of the ocean.

By adjusting the amounts of air and water, the crew can hit the point where the two forces are just equal. Then the sub floats freely, neither sinking nor rising in the water. To return to the surface, compressed air is blown into the tanks, forcing the water out and again reducing the weight of the sub until its density is less than that of water.

Ff **Fast Facts**

Submarines must be designed to withstand great pressure due to the weight of water above them. If you dive into a deep swimming pool, you feel discomfort at a depth of 10 feet because the pressure is 1.3 times as great as the pressure of the atmosphere. At 100 feet, a submarine experiences a pressure that is about 4 times atmospheric pressure, or 4 atmospheres. At 300 feet the pressure increases to almost 10 atmospheres.

How do hot air balloons and helium balloons rise in the air?

Ballooning is a popular pastime, so popular that some races have as many as a thousand contestants. How does the propane burner in the basket beneath a large balloon cause it to rise in the air?

Hot air balloons use the same relationships among pressure, volume, and temperature that we discussed concerning air pumps. As the burner rages beneath the air in the balloon's envelope, the temperature rises so the air molecules move faster. When the air gets hotter, its volume or its pressure must increase at the same time. Because the balloon is open on the bottom, air can escape freely, so the pressure remains the same as the pressure of the air around it. That means the volume of air increases. As the volume increases, the balloon inflates, until it reaches its maximum size. After that, any increase in volume forces air out of the balloon through the opening.

As air is forced out of the balloon, the mass of air that it holds decreases while the volume stays the same. That means the density of the balloon decreases. Because the principles of buoyancy apply to all fluids, whether they are liquid or gas, the change in the balloon's density has the same effect as the change in density of a submarine in water. If enough air is forced out, the average density of the balloon, its basket, and the people in the basket becomes less than the density of the air surrounding it. Then the balloon rises due to the upward force of buoyancy. To lower the balloon, the pilot opens a flap at the top. Warm air escapes through the top and is replaced by denser, cooler air from the surrounding atmosphere.

 Uncommon Sense

It is sometimes stated that a hot air balloon rises because heat rises. Heat, being a form of energy rather than a substance, cannot rise. The heated balloon rises because its density is less than that of the cooler air surrounding it. If you heat air in a closed container so that its volume cannot change, the container will not rise in the air. However, the pressure inside the container will increase.

A helium balloon also depends on buoyancy to rise into the atmosphere. Helium is much less dense than air because its particles have smaller mass than the particles of nitrogen and oxygen. One liter of helium has a mass of 0.179 grams while 1 liter of air has a mass of 1.25 grams. So if you have a 1-gram balloon filled with helium, its mass is less than the mass of air that it displaces. In a vacuum, it would read 1.179 grams when placed on a scale, but in air, it rises above the scale.

The balloon rises because the buoyant force is greater than the force of gravity that pulls it down. If you have enough helium displacing air, the upward force can lift a large mass. Think about the blimps that carry camera crews above sporting events.

Ff ❚ Fast Facts ❱

Like any other substance, helium has mass and is attracted to Earth by the force of gravity. Helium will only rise if it is surrounded by something (such as nitrogen and oxygen) with a greater density. If all the air were removed from a room so that there is a complete vacuum, a helium balloon and a bowling ball would fall from a shelf to the floor at exactly the same rate.

The blimp itself is very heavy and it is capable of lifting a large load. The difference in the mass of helium that fills the blimp and the mass of the air that would fill the same space provides all the lift.

What happens if the helium leaks from the balloon? You have no doubt seen the result. As the helium leaks through the wall of the balloon or around the closure, the balloon shrinks. The volume changes but the mass of the balloon does not change much at all because most of the mass is contributed by the balloon material, not the helium. As a result, the density of the balloon gradually increases. When the pull of gravity is stronger than the upward push of the air, the balloon sinks to the floor.

How can an airplane that weighs many tons stay in the air?

The difference in density of a helium balloon and the air around it can explain the rise of a balloon. An airplane, however, is a different story. You know from experience that the metal components of the airplane, the passengers, and the cargo have a density that is much greater than air. Did you ever look down at the ground, 5 miles below, and wonder just what it is that is keeping you in the air?

The basic reason that an airplane can fly is the same as the reason a balloon stays aloft—the force pushing it upward is equal to the force pulling it down. The biggest difference is that an airplane is much denser than air, so it is not buoyed up by a static difference in air pressure. The lift of an airplane comes from moving air. According to *Bernoulli's principle*, the pressure of a fluid decreases as it moves faster.

De Definition

Bernoulli's principle states that the pressure exerted by a fluid decreases as its velocity increases.

If you look closely at an airplane wing, you will see that it is curved on top and flat on the bottom. As the air flows past the moving wing, the air above the wing has to move farther because of the curvature. In order to do that, the air on top must move faster and therefore have a lower pressure. The pressure below the wing is higher, so it pushes the plane upward.

Most people have investigated the Bernoulli principle without realizing it. Have you ever held your hand outside the open window of a moving car? If you hold it perfectly flat, you feel pressure along the front edge but no force up or down. Angle your hand slightly upward, though, and you feel a strong lift that tries to push your hand up and away.

Another common illustration of the Bernoulli principle requires a sheet of paper and a puff of air. Hold the paper in front of your mouth and blow above it. You might think the moving air would force the paper downward, but you observe just the opposite. The moving air above the paper exerts less force than the stationary air below it and the paper moves upward.

Ff Fast Facts

A baseball pitcher uses Bernoulli's principle to throw a curve ball. The ball is launched with a strong spin so that one side of the ball has a greater velocity toward the batter than the other side. That means that on one side the air is moving faster relative to the ball and therefore exerts a lower pressure. On the other side the air is moving slower, exerting a higher pressure. The ball is pushed toward the lower-pressure side, causing its path to curve.

Watch what happens to the shower curtain the next time you turn on the water. The curtain immediately moves inward. The falling water from the shower head carries air along with it. The lower pressure on the inside of the shower curtain causes it to move inward—yet another example of the Bernoulli principle.

Why do your ears pop during takeoff and landing?

If you have flown in a large jetliner, you likely experienced a "popping" in your ears during the takeoff and landing parts of the flight. Sometimes the same thing happens as you drive in the mountains. What causes your ears to become uncomfortable and then pop as you travel?

The popping comes as a result of changes as you travel through a fluid—air. The air in the atmosphere exerts a downward pressure on everything below it, including the gases that make up the lower atmosphere. That means the pressure decreases as you go upward through the air and increases as you go downward. In general, you don't feel these changes inside your body. However, your inner ear is filled with air, a gas.

Ff **Fast Facts**

You can usually help your Eustachian tubes in the job of equalizing pressure by yawning, chewing, or other motions that open and close the tubes. Doctors do not recommend closing your nose and mouth and blowing, however, because you risk forcing fluids into your ears or damaging the eardrum.

The moving parts of your ear, especially the eardrum, need for the pressure inside and outside the ear to be fairly close to the same in order to work correctly. Small tubes, called the Eustachian tubes, provide an opening to pump air into or out of the inner workings. Under normal conditions, air pressure changes very gradually and the Eustachian tubes do their job without your knowledge.

An airplane moves up and down through the atmosphere very rapidly, however, so the tiny tubes can't always keep up with the change. As you ascend, air becomes trapped in the inner ear, pushing your eardrum outward. You feel discomfort and have trouble hearing because the eardrum can't vibrate normally. When the Eustachian tubes open, the air escapes rapidly. You hear a popping sound and feel the pressure inside and outside your ear become equal. The opposite effect occurs on the way down. Air pressure outside your ear increases, pushing the eardrum inward. The pop comes as air rushes through the tubes to your inner ear.

The cabins of airplanes that fly at high altitudes are normally pressurized for the comfort of passengers, but the difference in pressure between the cabin and the outside is not enough to cause the effects often seen in movies. A small crack in the window or hole in the structure will allow air to escape but it will not suck passengers to their deaths. Opening a doorway is still not recommended, however, due to the risk of falling through the large opening.

Does quicksand really pull people under?

Many older movies set in the American West or in the jungle include a quicksand scene. Someone falls into quicksand and is slowly pulled under. The evil victim disappears beneath the surface while the good victim is rescued in the nick of time when our hero drops a vine from above and pulls the victim straight up and out. But does quicksand really pull people beneath its surface?

You may not think of quicksand as a fluid because sand itself is a solid. Quicksand is ordinary sand in which a large amount of water reduces the friction between particles and allows them to move easily past one another. It is this ability of the particles to move that makes quicksand a fluid with properties similar to a liquid.

Quicksand occurs when sand or sandy soil becomes saturated with water, making a batterlike mixture. Generally, there must also be some kind of agitation, such as water flowing from a spring. Earthquakes can also provide the agitation needed to turn very wet sand into quicksand, sometimes causing buildings to sink into the ground.

If you happen to fall into quicksand, all is not lost. In the movies, quicksand pulls people under. In real life, the density of quicksand is greater than the density of the human body. Based on the principle

Ff **Fast Facts**

The "quick" in quicksand does not refer to motion. It comes from an old usage of quick as "living," because quicksand seems more alive than regular sand. Quicksand can occur anyplace at which there is a source of moving water beneath a layer of sand or gravel. Normally it is no more than 3 or 4 feet deep.

of buoyancy, that means you will float in quicksand. Normally you will sink only as deep as your waist. But getting out is a bit more complicated than grabbing the vine that your hero dangles. If you place pressure against the wet sand, it tends to lose water and become more solid. Pull straight upward and it will feel like your foot is in concrete.

The experts advise that the first step in getting free comes straight from *The Hitch-hiker's Guide to the Galaxy:* don't panic. Then slowly and carefully get rid of anything that adds weight and density to your body. Wiggle your legs slowly to get more water around them. Finally move toward the edge with a slow, gentle swimming motion.

> "Let us hope that the advent of a successful flying machine, now only dimly foreseen and nevertheless thought to be possible, will bring nothing but good into the world; that it shall abridge distance, make all parts of the globe accessible, bring men into closer relation with each other, advance civilization, and hasten the promised era in which there shall be nothing but peace and goodwill among all men."
>
> —*Octave Chanute (1832–1910)*

Chapter 6

Chemistry—Matter

"You will die but the carbon will not; its career does not end with you. It will return to the soil, and there a plant may take it up again in time, sending it once more on a cycle of plant and animal life."

—Jacob Bronowski (1908–1974)

Chemists study matter, which is defined as anything that has mass and volume. Everything around you is made of matter. Surprisingly, everything is composed of only about a hundred different basic types of matter, called elements. Some elements are quite familiar: iron, gold, oxygen, carbon, lead, and helium. You may have never heard of others: osmium, hafnium, and gadolinium. A few, such as the americium used in smoke detectors, do not occur naturally. These elements are produced in huge laboratories. It may seem strange that everything around you can be built with only 100 types of building blocks. Keep in mind, though, that the English language, with all its possibilities and nuance, is built of 26 letters.

Each element has a smallest particle, called an atom. Although atoms are themselves made of smaller particles, an atom is the smallest unit that has the properties of the element. The basic particles of atoms are protons, neutrons, and electrons. You have probably seen pictures showing an atom as a nucleus with electrons orbiting around it. Although this model does not exactly match our current understanding of atomic structure, it is close enough to be useful.

The nucleus of an atom has protons and neutrons, which make up most of its mass. Around the nucleus, electrons are in constant motion, making up most of the atom's volume. Altogether, they build a structure that is unimaginably small. Ten million atoms could make a line across the period at the end of this sentence. Even so, the properties of all matter depend on the interaction of atoms, one at a time.

Why is it more important to paint steel than aluminum?

Rust is a serious concern to anyone who builds or uses objects made of iron or steel. Take a look at an old car that has been sitting in a junkyard for a couple of decades. Although the shell of the vehicle still has its original shape, you can push your hand through the rusted steel with very little effort. What happens to a metal object when it rusts?

Rust is a *chemical compound* that forms when electrons are transferred from iron atoms to oxygen atoms. The two elements combine in a ratio of two parts iron to three parts oxygen to form ferric oxide, or rust, which has properties that are very different from those of either element alone. In fact, the iron ore that we find in nature is essentially rust, which explains why water in an iron mine has a characteristic red color.

De **Definition**

A **chemical compound** is a substance formed by the combination of two or more elements in a fixed proportion of atoms. The atoms are bound together by transfer or sharing of electrons, making a substance whose properties can be very different from those of the elements of which it is composed.

Generally, three things must come together to cause substantial rusting of steel—iron, oxygen, and water. Electrons are not transferred very efficiently from the iron atoms to oxygen as it occurs in air. However, water molecules very effectively aid in the movement of electrons between the two elements. That is one reason that cars tend to survive longer in the dry deserts of the Southwest than in places with humid climates.

Paint prevents rust by keeping oxygen and water away from the iron atoms. The modern paints used on automobiles essentially form a hard layer of plastic that bonds strongly to the metal surface. Oxygen cannot penetrate the paint unless it becomes scratched or chipped. A small break in the coating, however, can open the metal underneath to corrosion.

Although aluminum is often painted for decorative purposes, the coating is not necessary for surface protection. It's not that aluminum does not react with oxygen. In fact, aluminum reacts with oxygen even more readily than iron does. The difference is that when iron rusts, the soft iron oxide flakes away, revealing more iron atoms, while aluminum oxide forms a tightly bound coating on the aluminum surface. The aluminum oxide forms a protective layer that cannot be penetrated by oxygen and water.

Fast Facts

Stainless steel is an alloy of iron that does not rust readily. In addition to iron, stainless steel contains at least 12 percent chromium and often other metals as well. The chromium atoms on the surface react with oxygen to form a protective layer that prevents further reaction by atoms beneath the surface.

Why do snowflakes have six sides or points?

If you look closely at a snowflake, you will usually find that it has six sides or points. A snowflake is a crystal formed by the arrangement of molecules of water in its solid form. Snow crystals form when water begins to freeze around small particles of dust inside clouds. Snow forms in several shapes depending on the temperature: near 32°F, the crystals form small flat plates; a few degrees cooler and they take on a needle or pencil shape; large, lacy shapes form at about 5°F or colder. No matter which shape the crystals take, they have six sides.

Many solid materials in nature form crystals when liquid or gases cool to become solid. The smallest particles of the material—atoms or molecules—are constantly in motion. As these particles cool, they lose energy and move more slowly. Eventually their energy of motion drops to a level at which the forces that cause particles to attract one another are stronger than the forces that cause them to move apart. Then a solid crystal forms. The shape of the crystal is determined by the forces that cause particles to stick together.

Molecules of water are made of two atoms of hydrogen bound to an atom of oxygen, forming the shape of a V. The oxygen and hydrogen atoms share electrons, forming a chemical bond, but the electrons are attracted to the oxygen atom a bit more than they are attracted to the hydrogen atom. As a result, the point of the V has a bit of a negative charge and the tips have a bit of a positive charge. Because of these charges, water molecules are attracted to one another. As the solid forms, the oxygen atom of one molecule lines up close to the hydrogen atoms of two other molecules. The most stable arrangement occurs when six water molecules form a ring of V shapes. This is the beginning of a six-sided crystal.

The crystal grows as more and more water molecules join the original group. The attractions between the charged parts of the molecules cause the crystal to grow in six equally spaced directions. Eventually, after many trillions of molecules have joined together, the crystal begins to fall. As it descends, the snowflake can gain or lose molecules, so the shape is constantly changing. Differences in temperature and humidity cause each pattern to change as it moves up and down through the atmosphere, so two snowflakes falling close to one another may have very different shapes. In general, though, the six sides remain.

Why does dry ice not melt like regular ice?

Imagine ordering a frozen-food item for shipment from a company across the country. If the item is packaged with a block of ice, it may stay cold until it reaches you, but even with good insulation, your product is likely to arrive in a pool of water. Instead of ice, cold products are generally cooled during shipping by dry ice—solid carbon dioxide. When the package arrives, the block of dry ice is much smaller than when it was shipped but there is no pool of liquid.

At atmospheric pressure, carbon dioxide is a solid at –109°F (–78.5°C), cold enough to keep water-based foods frozen solid. The major advantage of shipping with a block of dry ice is that it goes directly from solid to gas in a process called *sublimation*. This process is sometimes used to create fog for stage performances.

If dry ice is placed in warm water, it quickly sublimates and the gas flows away, carrying with it tiny droplets of water, making a fog. Because the gas is cold and carbon dioxide is heavier than air, the fog flows across the stage instead of rising above it. This is very useful for creating an eerie feel in a performance of *Macbeth*. Dry ice is also used in a process, similar to sandblasting, to clean surfaces. Solid pellets are sprayed at the surface, stripping contaminants and paint. The pellets sublimate into a gas and don't leave behind piles of sand to be cleaned up.

Why doesn't carbon dioxide form a liquid like most substances? In fact, it does, but only under pressure. The state—gas, liquid, or solid—of a substance depends on temperature and pressure. For example, because atmospheric pressure is lower at high altitudes, the boiling point of water is several degrees lower in Denver than in Houston. If you increase the pressure to about five times normal atmospheric pressure, carbon dioxide becomes a liquid at room temperature.

Carbon dioxide fire extinguishers hold liquid carbon dioxide under pressure. When it is released from the extinguisher, the carbon dioxide expands rapidly, which causes it to cool. A spray of carbon dioxide "snow" rapidly sublimates to form a gas that smothers the flame. Liquid carbon dioxide can also be used in some dry cleaning processes for clothing.

De **Definition**

Sublimation is the process in which a material passes directly from the solid phase into the gas phase without becoming a liquid.

Ff **Fast Facts**

Under the right conditions, water sublimates. Below 32°F (0°C) water exists as a solid at atmospheric pressure. Although it cannot melt at colder temperatures, some ice passes directly to the atmosphere as a gas. You can see this process when frost disappears from the outside of a window on a very cold day. In parts of Antarctica, cold winds cause several centimeters of ice to sublimate from the surface each winter.

How do synthetic diamonds differ from real diamonds?

Diamond is the hardest naturally occurring substance, which makes it useful for industrial cutting tools. It also has an amazing ability to play with light, which makes it useful for adornment. However, diamonds are relatively rare and expensive to mine. That makes them expensive to buy.

A number of diamond substitutes or simulated diamond, including cubic zirconia and moissanite, fill a niche for low-cost diamond look-alikes, but they are not real diamonds. These materials have a different chemical composition and are easily identified by an expert, or frequently by the casual observer. For industrial applications, simulated diamond is not at all suitable.

Ff | **Fast Facts**

Natural diamond can only form at the high temperatures and pressures found at least 100 miles beneath the surface of Earth, and they are carried to the crust by deep volcanoes. Most natural diamonds were formed between 1 billion and 3.3 billion years ago and carried to the surface much later.

Since the discovery in 1797 that diamonds are made of pure carbon, people have looked for ways to manufacture a real diamond. A hundred miles beneath Earth's surface, extreme pressures and temperatures squeeze carbon atoms into a network in which they are tightly bound to one another. Until recently, there was no way to recreate these conditions on the surface (with the possible exception of Superman's fist). The first synthetic diamond was made by General Electric in 1954. Recent developments have begun to challenge nature's monopoly on diamond making, and not incidentally, the profits of DeBeers, the largest producer of mined diamonds.

Synthetic diamonds are real diamonds, having the same chemical composition as natural mined diamonds. They are created in hours rather than millions of years. Every year, manufacturers produce more than 100 tons of synthetic diamonds. In general, they have colors that make them less than ideal as gems but they are very important for industrial applications. Synthetic diamonds have been used in cutting tools, surgical instruments, and electronics applications.

What is the most abundant element?

When you think of elements, some, such as oxygen, carbon, and iron, seem to be very common and abundant. Other elements, such as gold, silver, and platinum, are much rarer and very expensive. Which of the 100 or so elements are the most abundant?

The answer to that question depends on where you look. If you are interested only in elements that are readily available here on the surface of Earth, then the most abundant element is oxygen. Most of the mass of the ocean's water is made of oxygen and almost all of the minerals that make up the rocks and soil of Earth's crust contain a large amount of oxygen. The following table shows the eight most abundant elements by mass in the crust of Earth, which make up 98 percent of the total.

Most Common Elements in Earth's Crust

Element	Weight Percent
Oxygen	46.1
Silicon	28.2
Aluminum	8.23
Iron	5.63
Calcium	4.15
Sodium	2.36
Magnesium	2.33
Potassium	2.09

It may seem surprising that silicon is the second most abundant element. It is only with the introduction of solid-state electronics that most people have even become aware of silicon as a substance. It has been used for a long time, however. Glass—and the sand that glass is made from—is composed mainly of silicon dioxide. Also surprising is the absence of some well-known elements from the list. Carbon and nitrogen are found in a great many common compounds, but they are only a fraction of the remaining 2 percent of the matter around us. Hydrogen is also missing, which is something of a surprise because it is actually a very abundant element in the universe, as we will see shortly. If you look at Earth as a whole, not just the crust, then the most abundant element is iron, which makes up most of the core of the planet.

However, if you consider not just Earth, but the entire universe, the picture is very different. While eight elements combine to make up about 99 percent of Earth's crust, only two elements make up 98 percent of the universe—and neither of them is on the previous list. The following table shows the relative abundances of the top eight elements in the universe. As you can see, it is mostly hydrogen, the simplest element.

Most Common Elements in the Universe

Element	Weight Percent
Hydrogen	73.9
Helium	24.0
Oxygen	1.07
Carbon	0.46
Neon	0.13
Iron	0.11
Nitrogen	0.09
Silicon	0.06

An interesting aspect of modern models of the universe is that all of the elements combined make up only 4 percent of the universe as a whole. The rest is called "dark energy" and "dark matter." However, neither of these has ever been detected, so no one knows their properties.

Is jet fuel more dangerous than gasoline?

When an airplane must land with malfunctioning landing gear, dozens of fire trucks and other emergency vehicles stand ready to move to its final destination. Sometimes the crashed plane is quickly enveloped in flames. Is jet fuel an inherently dangerous chemical, more hazardous to handle than the gasoline that you have in your shed for your lawnmower?

While jet fuel is a flammable compound, it is actually less likely to catch fire or explode than ordinary gasoline. Most jet fuel, like gasoline, is made of hydrocarbon compounds that are distilled from crude oil. These molecules consist of chains of carbon atoms linked to one another and to hydrogen atoms. When the hydrocarbon burns, it reacts with oxygen in the air to make carbon dioxide and water—and it releases a lot of energy. The fewer carbon atoms in the hydrocarbon, the easier it catches fire. Natural gas, or methane, has a single carbon atom in each molecule and burns very easily. Gasoline is a mixture of hydrocarbons with varying carbon chain lengths, averaging seven or eight carbons per molecule. Jet fuels, such as kerosene, have more carbons per molecule than gasoline so they do not ignite as easily.

One measure of flammability is the temperature at which a mixture of a vapor and air will ignite when exposed to an energy source, such as a spark. For gasoline, the flash-point is approximately –40°F. For jet fuel, it is +100°F. An open container of gasoline is much more hazardous than an open container of jet fuel because the vapors above the jet fuel will not burn below 100°F. However, when a jet airplane undergoes a crash landing, there is still a significant risk of fire. The engines are hot and friction between the runway and the body of the plane also generates a lot of heat. That's why the firefighters are standing by.

The low flashpoint of gasoline is an important reason to take care when filling the tank of a hot lawn mower engine. The heat of the engine rapidly vaporizes gasoline that is splashed on it. The vapor above the engine can ignite at any temperature above –40°F, which includes just about all the conditions in which you would be mowing your lawn. We are so used to using gasoline, it is easy to forget that it can be a dangerous chemical.

Us | **Uncommon Sense**

Watch a movie with a massive chase scene and you are almost certain to see at least one car explode when it crashes. Guess what—it doesn't happen in real life. An explosion can only occur when a mixture of gasoline vapor and air is in a narrow range. Most cars have very sturdy gas tanks that do not rupture in a collision, and even if they did, an explosion could not occur until a substantial amount of gas had leaked.

Why do apple slices turn brown?

Slice an apple into pieces and, depending on the variety, the flesh you see is pure white to light yellow or pink. Within a few minutes, however, the apple slices have started to turn brown. The browning occurs because enzymes in the apples start a chain of chemical reactions. The enzyme in fruits that causes browning is called polyphenol oxidase (PPO). When the enzyme and oxygen are exposed to compounds called phenols that are present in the tissues of apples and other fruits, the phenols are converted to other compounds called quinones. These quinones rapidly react with proteins in the apple to make the brown-colored compounds that you see on the surface of the apple slices.

If the PPO is inside the apple, why doesn't the flesh of the apple turn brown before it is sliced? The main reason is that the skin of the apple keeps oxygen outside. In addition, the PPO and the phenols are generally located in different cells inside the apple. As a knife ruptures the cells, the compounds mix. In fact, there are some cases where this mixing can occur without cutting the apple. Drop an apple on the floor and the impact ruptures cells beneath the surface. In a short time, a bruise develops under the skin. The brown bruise is caused by the same reaction, using oxygen that is available in the cells.

You can slow the browning of apple slices by keeping oxygen away from the surface if you cover the fresh slices with water, sugar, or syrup. A splash of lemon juice will also keep the slices fresh looking. Lemon juice, and the juice of other citrus fruits, contains compounds that are antioxidants. These compounds, including Vitamin C, react with oxygen very quickly, so the splash of juice keeps oxygen from reaching the PPO in the apple.

Ss Science Says

"About seven years later I was given a book about the periodic table of the elements. For the first time I saw the elegance of scientific theory and its predictive power."

—Sidney Altman (1939–)

PPO is present in many plants and in many foods that turn brown rapidly, including bananas, mushrooms, peaches, and pears. The purpose of PPO in these plants is not completely understood. The PPO may be part of the cycle by which plants use oxygen in their cells, or the PPO may provide protection from certain pests, or both.

How can lightweight body armor stop a bullet?

Whenever police or military personnel enter a situation that could involve a firefight, they put on body armor, sometimes called a "bullet-proof vest." This armor is a descendent of the suits of armor worn by knights in battle long ago, although its construction is very different. How can a vest made of woven cloth stop a bullet?

Think about what happens when a professional tennis player returns a 100 mph serve. The racquet consists of a mesh of interwoven filaments that form a net. When the ball and the face of the racquet connect, the force of the collision is divided among many separate strands. As each strand bends, it spreads the force over its whole length and transfers some of the force to each of the perpendicular strands of the weave.

If the entire force of the collision were concentrated on a single string, that string would likely break, but, by spreading the impact, the ball is stopped and then its motion is reversed.

Woven body armor works on the same principle. Because the bullet is smaller than a tennis ball and moving many times faster, the woven net must have a very tight weave. In addition, the strands must be made of a very strong, but flexible, material. The fabric of modern body armor is woven of synthetic fibers that are lightweight like cotton or silk, but have a *tensile strength* many times greater than steel fibers of the same size.

As each fiber in the vest bends, it absorbs energy and spreads it across the length of the fiber and across the length of the fibers connected to it. As a result, a dense net of fabric can absorb a lot of energy. Modern body armor consists of 20 to 40 layers of densely woven fibers, each layer absorbing and spreading part of the impact force.

These two functions—stopping the bullet and spreading the force of the impact—are the source of the armor's protection. The person wearing the vest will still feel the energy of the bullet's impact, but it is spread over a wide area of the body to prevent

De | **Definition**

Tensile strength is a measure of the stress required to stretch a material until it breaks. It is calculated by measuring the maximum load that the material can support and dividing that load by the original cross-section of the material. It is commonly expressed in units of pound per square inch.

serious injury. Some of the newest types of armor have a fluid material between the woven layers. This fluid actually solidifies when it is under stress. In this way, it is like a mixture of cornstarch and water which can be rolled into a ball as long as pressure is continually applied but becomes a liquid as soon as the pressure is removed. As the fluid solidifies in the vest, it can absorb a lot of energy from the moving bullet very rapidly and then release the energy slowly as it again becomes a liquid.

Ff | **Fast Facts**

The fiber with the strongest known tensile strength is spider silk. A spider's web is built of strings of protein that bend but don't break under extremely high forces. Researchers are working on ways to make artificial spider silk, which could be used to make body armor far superior to those made with materials available today.

"There is no better, there is no more open door by which you can enter the study of natural philosophy than by considering the physical phenomenon of a candle."

—Michael Faraday (1791–1867)

Chemistry–Matter on the Small Scale

"Matter, though divisible in an extreme degree, is nevertheless not infinitely divisible. That is, there must be some point beyond which we cannot go in the division of matter. The existence of these ultimate particles of matter can scarcely be doubted, though they are probably much too small ever to be exhibited by microscopic improvements. I have chosen the word atom to signify these ultimate particles."

—John Dalton (1766–1844)

The nature of all the matter around you depends on the atoms of which it is made. Solid materials are solid due to strong interactions between atoms of the material. Hot materials are hot because the atoms are moving very rapidly. Some metals carry an electric current easily because their atoms lose electrons easily. Every property of matter can be explained through an understanding of the small particles of which all matter is made.

Why does spreading salt keep a road free of ice?

If you live in a place with freezing winters, at some point you have probably followed a truck spreading salt on the road. Shortly after the truck passes, ice on the road melts, making it safer to drive. Why would putting salt on top of ice make it melt?

Actually, the salt does not really melt the ice. Forces between molecules of water cause them to be attracted to one another. As a substance, such as water, cools, its molecules move more slowly because they have less energy. If the energy of the molecules is less than the forces that cause them to be attracted to one another, they cling together, forming a solid. Ice forms at 32°F because that is the temperature at which the water molecules form a solid crystal. However, not every molecule of water is tightly linked. Some are constantly losing energy and freezing while others are gaining energy and melting. This process occurs continually at the surface of a piece of ice. If water freezes and ice melts at the same rate, the ice appears to be unchanging.

When salt is added to the top of the ice, it dissolves and its particles don't fit neatly into the crystal pattern at the surface. The salt tends to prevent the freezing of the water. However, the salt does not affect the melting of ice, which continues at the same rate. Because the melting rate is now faster than the freezing rate, the amount of ice decreases and the amount of water increases. It appears as though the salt caused the ice to melt.

Ff **Fast Facts**

Salt is not the only substance that lowers the freezing point of water. In fact, anything that can be dissolved in water will have the same effect. When you add ethylene glycol to the radiator of your car, it lowers the freezing point of the water in the radiator, preventing the damage that would occur from running the engine without coolant if the water were all frozen in the radiator.

Eventually, if the mixture of salt and water gets cold enough, it will freeze, so the effect of the salt is a lowering of the freezing point of water. This is an important consideration. On a really cold night, salting the roads does not help because the saltwater still loses enough energy to slow down and freeze. Below 2°F, even a solution with 20 percent salt will freeze. At very low temperatures, the trucks spread sand or gravel, which improves traction on top of the ice without melting it.

The freezing point change does not vary based on what material dissolves in water. Instead, it depends on the number of particles dissolved in a specific volume of water. Ordinary salt, sodium chloride, dissolves to form two particles, called *ions*, for each unit of salt. A different compound, calcium chloride, is often used on ice. When calcium chloride dissolves, it forms three ions, so a smaller amount is needed to obtain the same effect. In addition, calcium chloride is less likely to damage concrete surfaces and living plants.

> **De** Definition
>
> An **ion** is an atom that has an electric charge due to the gain or loss of an electron. If the atom gains an electron, its ion has a negative charge. If it loses an electron, its ion has a positive charge.

Why does a rug feel warmer than a tile floor?

Walking barefoot across a bathroom on a cool night, you will notice a big difference in the feel of a tile floor and a rug on top of it. While the tile feels cold against your feet, the rug is quite comfortable. You know that the temperature of both surfaces should be the same as the temperature of the air around them, so why does one feel warm and the other cool?

A thermometer will show that rug and floor are indeed at the same temperature. However, our sensation of how hot or cold something is depends on more than just its temperature. Heat is actually a transfer of energy that always flows from a higher temperature to a lower temperature. An object feels cold when energy is transferred as heat from your skin to the object.

Although the difference in temperature between your skin and the floor is the same as the difference in temperature between your skin and the rug, the transfer of heat is not the same. The two materials have a different *thermal conductivity*.

> **De** Definition
>
> The **thermal conductivity** of a material is its ability to transfer heat by the collision of atoms and molecules. Materials such as metals, in which the atoms are tightly packed, generally have a higher thermal conductivity than materials such as air, in which the particles are farther apart.

Every atom and every molecule is in motion all the time—moving around and bumping into nearby atoms. When heat flows into an object, its temperature increases, which means that these particles move faster. As these atoms and molecules move, they collide with others. Each collision transfers energy, again as heat. Materials in which the particles collide very frequently transfer heat throughout the object more efficiently than materials in which there are fewer collisions. A greater number of collisions means a higher thermal conductivity, because heat is rapidly carried away through the material.

In order to understand the importance of thermal conductivity in everyday things, think about the difference in the effect of air temperature and water temperature on your body. On an 80°F day, you jump into a swimming pool whose water is also at 80°F. The air feels warm but the water feels cool. Why? Air has a very low thermal conductivity, so very little heat is transferred from your skin to the air. Water has a much higher thermal conductivity, so you feel cold when you jump into the water.

The low thermal conductivity of air is one reason that the carpet does not feel cold. The fibers of the carpet have a lower thermal conductivity than the ceramic of the tile. The greatest obstruction to heat flow, however, is the air that is trapped within the pile of the carpet.

Ss **Science Says**

"If, in some cataclysm, all of scientific knowledge were to be destroyed, and only one sentence passed on to the next generations of creatures, what statement would contain the most information in the fewest words? I believe it is the atomic hypothesis (or the atomic fact, if you wish to call it that) that all things are made of atoms—little particles that move around in perpetual motion, attracting each other when they are a little distance apart, but repelling upon being squeezed into one another."

—Richard P. Feynman (1918–1988)

How can diamond and graphite both be pure carbon?

Most pure elements occur in one form that has particular properties that are always the same. Pure gold is always a shiny, yellow metal; nitrogen is a colorless, odorless gas; chlorine is a corrosive, greenish gas. How can diamond and graphite, which have such different properties, both be forms of pure carbon?

The difference in types of carbon is due to the arrangement of atoms in large crystals or sheets. Because a carbon atom is able to make as many as four chemical bonds with other atoms, carbon atoms can join together in large arrays.

The difference in the properties of the forms of carbon is built into the way the atoms are arranged. Graphite consists of broad, flat sheets in which each atom forms bonds with three other carbon atoms. In a sample of graphite, these sheets are stacked, one on top of another. A piece of graphite that is 1 millimeter thick consists of about 30,000,000 layers. The forces that hold these layers together are very weak compared to chemical bonds, so the layers tend to slip and slide past one another. As a result, graphite has a greasy feel and makes a good lubricant.

Uncommon Sense

Pencil "lead" does not contain the metal lead at all. It is a mixture of graphite and clay. Graphite was originally called lead because the people who first discovered natural deposits, in the 1500s, thought that they had discovered a lead deposit. The name *graphite* comes from the Greek word meaning "to write."

In another form, carbon is a powdery black solid, known as lampblack, which is commonly used in inks or a chunky solid—charcoal. On an atomic scale, lampblack and charcoal resemble small sheets of graphite that are randomly arranged rather than layered.

In a diamond, on the other hand, carbon atoms form bonds with four other atoms in a tetrahedral pattern. The bonds between atoms are very strong and do not break easily. The strength of these bonds makes diamond the hardest known natural substance.

Fast Facts

Carbon is not the only element that occurs in several different forms. When two oxygen atoms are bound together as a single molecule, which we also call oxygen, they are an essential substance for humans. When three oxygen atoms are bound together to form ozone, they make a toxic gas.

A fourth form of carbon was discovered in 1985. In this form, carbon is arranged in large molecules that look something like soccer balls. Because a model of the first such molecule to be discovered—a 60-carbon-atom ball—resembled a geodesic dome, the molecule was named buckminsterfullerene after the architect who designed that type of dome. A number of carbon molecules have been discovered since and, as a class, they are known as fullerenes. Scientists who study fullerenes hope to use some of their unusual magnetic and electrical properties and to trap other atoms or molecules inside the molecule in order to obtain new properties.

How do soaps and detergents work?

Try washing your children's clothes after they've played outside all day and eaten peanut butter sandwiches—without using any detergent. It is reasonable to predict the shirts will come out of the washer in something less than perfectly clean condition. And what about washing those dirty hands without any soap? How do soaps and detergents make washing more effective? And what is the difference between soap and detergent, anyway?

To start, let's take a look at why dirt sticks to us, our clothes, and everything else that gets dirty. Sometimes, particles of dust just sit on a surface. A good dust cloth will take them off. Then there is the stuff that sticks—won't rub off, shake off, or rinse off with pure water. This dirt is the real challenge, because it is held in place by oil or grease—things that seem to cling tightly to skin, cloth, and even hard surfaces like dishes. Water rolls away from oil or grease with no effect.

Molecules of water are polar, which means that they have small electrical charges— positive on one end of the molecule and negative on the other end. Oil molecules, made of carbon and hydrogen, do not have electrically charged ends. Electrons in the bonds that make up the molecules are attracted just about equally by the two atoms involved in the bond. So, if you put water and oil (or grease, which is basically made of great big oil molecules) together, the water molecules congregate in one place and the oil molecules stick together elsewhere. You can remove oil with a nonpolar solvent, such as paint thinner, alcohol, or kerosene, but these are not usually compatible with things like skin and electrical appliances.

This is where soap comes in. Soap molecules have an interesting mix of polar and nonpolar parts. One end of the soap molecule has a bond between an oxygen atom and a metal atom, usually sodium or potassium. This bond is very polar, and when the soap molecule is placed in water it forms ions. The metal ion wanders off and interacts with water molecules while the negatively charged part of the soap molecule attracts the positive ends of water molecules and dissolves in the water. The rest of the molecule is made up of a long string of carbon atoms with hydrogen atoms attached to the chain. This sounds like the description of the oil molecules—and it is. Whenever one of these long nonpolar strands meets an oil molecule, they immediately latch onto one another.

The end result of these attractions is that the oil-like ends of the soap molecule are attracted to oil and grease. Soap molecules surround the nonpolar parts of dirt, with their polar ends sticking out in every direction. These polar ends are dissolved in the water, so the whole ball of dirt and soap leaves the fabric or your skin and, along with the (now grungy) water, heads down the drain.

Us | **Uncommon Sense**

Dry cleaning of clothing, because of its name, is often assumed to use a process without solvents. Actually, a dry cleaner removes greasy or oily dirt by tumbling clothes in a nonpolar liquid solvent which dissolves the grease and oil. It is called dry because no water is used. Because the solvent matches the dirt in polarity, detergents are not needed.

Soap is made by reacting animal or plant fats with an alkali, such as lye. Detergents are similar to soap in that they have a long nonpolar chain and a very polar end, but they are generally made from petroleum products. For some purposes, though, detergents work better than soap. Some water systems provide "hard" water, which means the water contains high levels of calcium or magnesium ions. When soap is added to hard water, these ions react with the soap to make a solid *precipitate*, which is then not able to dissolve dirt in water. This solid can be seen as a bathtub ring. Detergents do not form a precipitate, so they can clean in hard water as well as soft water.

De | **Definition**

A **precipitate** is an insoluble compound that forms when two or more substances in a solution react with one another. The precipitate does not remain in the solution.

How are different colors produced in fireworks?

Everyone loves a fireworks show on the Fourth of July. The loud booms of the exploding rockets accompany bright flashes of light. The colors are an important part of the display. You just know that at some point in the show you will see red, white, and blue flashes flying across the sky. How do fireworks manufacturers design their rockets to show the desired colors?

Each time one of the fireworks explodes, a carefully designed mixture of explosives, combustible fuel, and compounds containing metals provides the desired effect. The key to fireworks color lies inside the atoms of the metal atoms added to the rocket. The colors are produced by heating the metal atoms, which then emit light in a color that is characteristic of the metal.

Every atom consists of a nucleus surrounded by rapidly moving electrons. The energy of the electrons is not random, though. Every type of atom has a unique set of possible energy levels for its electrons. Under normal conditions, the electrons occupy the lowest allowed energy levels. The atom is then in its ground state. If an atom absorbs enough energy, in this case in the form of heat, the added energy moves one or more of its electrons to a higher level. When electrons are at higher energy levels than the ground state, the atom is in an excited state.

Atoms don't remain excited for very long. The electron quickly sheds the extra energy to get back to its ground state. The move from a higher energy level to a lower energy level is always accompanied by the release of a unit of energy as light.

De Definition

An **emission spectrum** is the electromagnetic radiation emitted by the atoms of an element when they are heated. Each element has a characteristic emission spectrum that can be used to identify that element.

For any given atom, there are many clearly defined energy changes that can occur when electrons lose energy. One specific wavelength (or color) of light is always emitted with each change. Using a prism, the colors produced by each element can be separated to make an *emission spectrum*. Emission spectra can be used to identify the elements in a chemical sample. While most of the changes release light in wavelengths that we cannot see, some energy changes produce light that falls into the visible spectrum of light.

Pure metals are used for a few effects. For example, burning magnesium metal produces a brilliant white light. For most colors, however, the fireworks designer uses a salt that includes the metal. What metals produce the colors of fireworks? Here are some of the common fireworks colors and the metals added to produce them:

Color	Metal(s)
red	lithium or strontium
orange	calcium
yellow	sodium
green	barium, thallium, zinc
blue	copper, lead
purple	cesium, rubidium
white	magnesium
silver	aluminum or titanium
gold	iron

You can investigate some of these effects in your own fireplace. Try tossing a few crystals of table salt (sodium chloride) into the flames. You will immediately see a brilliant yellow color in the flames as the sodium atoms become excited and then relax to the ground state. Or you can try using a sodium-free salt substitute, potassium chloride, which will make a reddish-purple colored flame.

Ff **Fast Facts**

The element helium was discovered in the sun before it was discovered on Earth. In 1868, astronomers looked at a spectrum of the sun during an eclipse. They noticed light at wavelengths which did not correspond to the emission spectrum of any element known at that time and concluded that they had observed a new element. The element was named helium based on the Greek name for the sun, *helios*.

What causes popcorn to pop?

You toss a bag with a few hard, yellow kernels into the microwave, turn on the power, and wait about three minutes. When you pull the bag out of the microwave, it is completely full of fluffy, white popcorn. Each popped kernel has expanded to about 50 times its original volume. How does something change so drastically with just a bit of added energy from the oven?

Like so many other things, you have to look at the tiny particles—the molecules of water and starch—to figure out what happens when popcorn changes. One clue to what is going on is that not just any corn will pop. There has to be just the right amount of moisture in the kernel or nothing happens. Another clue is in the kernels that remain hard and unpopped. A close look will reveal that many of them are cracked or otherwise damaged.

Ff **Fast Facts**

Popping corn is not a recent idea. Archaeologists have discovered unpopped (and sometimes still viable) kernels in many Native American sites. The oldest known popcorn was harvested more than 5,000 years ago.

Three things are needed in a popcorn kernel to make it function correctly when heated. First, the kernel is filled with starch. Starch is made of long molecules that string together hundreds or thousands of small sugar molecules. The second component is a hard cellulose shell on the outside that is pretty much impervious even to small molecules like water. The third component is water in just the right amount.

Inside a good kernel of popcorn is water, accounting for about 14 percent of its total mass. When you turn on the heat, the water inside the kernel changes from its liquid form and becomes steam. The big starch molecules in the kernel surround thousands of tiny pockets of steam, miniature bubbles surrounded by strands of starch. As the starch cooks, it expands and as the steam gets hotter and hotter, it also expands. The cellulose shell around them does not change, though, so inside the shell the starch and steam build a big head of pressure. Finally, the pressure ruptures the outer shell with a loud pop. The little bubbles expand, the steam escapes, and left behind is cooked starch—its familiar crispy, fluffy, white blobs.

It sounds pretty simple, but a lot of things can go wrong. The water must heat to about 400°F very rapidly. If it heats too slowly the starch cooks without forming the bubbles and you get a small dense chunk of starch instead of a big puff. If there is too little water, the kernel never pops, but if there is too much, it breaks too soon without cooking the starch sufficiently, leaving a hard cracked kernel. Finally, the popcorn kernel needs a good, strong shell that has no flaws or cracks, which would let the steam escape without building pressure. If your supplier has been really careful, only about 4 percent of your kernels will fail. On the other hand, if the popcorn has not been processed correctly, you may find almost half of them staring at you from the bottom of the bowl.

Ss **Science Says**

"One of the wonders of this world is that objects so small can have such consequences: Any visible lump of matter—even the merest speck—contains more atoms than there are stars in our galaxy."

—P. W. Atkins (1940–)

How can carbon be used to find the age of an object?

Archaeologists frequently study the sites of ancient villages or encampments to learn more about the people who lived at that time. One major challenge, though, is to determine exactly when the village or camp was populated. If there are wooden tool handles, ashes from a campfire, or bones from an ancient meal, then scientists can analyze the carbon in the artifact to determine its age. How can the carbon in an object be used as a calendar?

When we discuss chemical reactions of atoms, we focus on the electrons that move around the nucleus. In order to find the age of an object, we must look past the electrons, into the nucleus itself. An atomic nucleus contains protons and neutrons. Although the number of protons is always the same for all the atoms of any given element, there is some variation in the number of neutrons. Some combinations of protons and neutrons are very stable, but other combinations change over time, a process known as *radioactivity.*

De **Definition**

Radioactivity is the spontaneous decay of the nucleus of an atom. Radioactivity is a process that rearranges the protons and neutrons into a more stable configuration. Normally, it occurs by emitting an alpha particle—two protons and two neutrons—or a beta particle—one electron. Elimination of an alpha particle reduces the number of protons by two, forming an atom of an element whose atomic number is two less than that of the original atom. Elimination of a beta particle is accompanied by the change of one neutron into a proton, increasing the atomic number by one.

As the nuclei of radioactive atoms decay, particles are thrown out of the nucleus and the number of protons changes. Radioactive materials are dangerous because these particles can damage the tissues of living things.

Most carbon atoms have six protons and six neutrons. This type of carbon is called carbon-12 for the total number of particles in the nucleus. Some carbon atoms, however, are carbon-14, a type, or isotope, of carbon that has six protons and eight neutrons. Carbon-14 is a radioactive isotope of carbon, which decays by emitting a beta particle and changing to nitrogen-14.

The key attribute of radioactivity, for dating purposes, is that it occurs at a predictable rate. If you start with a given amount of carbon-14, half the atoms will change to nitrogen over a period of 5,730 years. Half of the remaining carbon-14 atoms will decay during the next 5,730 years. This period is known as the half-life of carbon-14. The half-life of an isotope is a characteristic of that isotope. Some half-lives are a fraction of a second, while others are billions of years, but for any given isotope the half-life is constant.

Carbon-14 is formed in the atmosphere. Radiation from space converts a small amount of nitrogen into carbon-14 through a nuclear reaction. Because this is a continual process and the breakdown of carbon-14 is a continual process, the ratio of carbon-12 to carbon-14 atoms is fairly constant. In chemical reactions, the two isotopes act the same way. This means that plants, which use carbon dioxide from the atmosphere, will therefore have the same ratio of carbon isotopes. Animals that eat the plants will also have the same ratio of carbon atoms. This ratio stays the same during life because living things are constantly receiving new carbon from the environment.

Us **Uncommon Sense**

It might seem that radiocarbon dating could be used to find the age of any fossil—for instance, the fossils of dinosaurs. However, radiocarbon dating can be used only to find the age of objects up to about 50,000 years old. Beyond that age, there is not enough carbon-14 left for an accurate measurement.

Radiation dating, using isotopes of uranium with very long half-lives, has been used to find the age of rocks that are several billion years old.

When a plant or animal dies, however, it no longer takes in new carbon. The ratio of carbon-14 to carbon-12 begins to change as the carbon-14 decays and is not replaced by new carbon-14. In order to find out how long ago something was alive, scientists measure the ratio of the isotopes. If there is exactly half as much carbon-14 as expected, the object came from something that died 5,730 years ago, one half-life. If the ratio of carbon-14 is one fourth the expected amount, the object is two half-lives, or 11,460 years old.

Chapter 8

Chemistry–Interactions of Matter

"If the history of technology tells us anything, it is that the future lies in the world of the very small."

—Eric Cornell (1961–)

When you put a log on a fire, it burns and releases energy as heat and light, leaving behind only a small amount of ash. A steel pipe, on the other hand, gets hot but does not otherwise change in the fire. Mix some baking soda in water and nothing happens, but when you add some lemon juice to the mixture, bubbles form and it foams out of the container.

Every substance has different properties that are defined by its interactions with other substances and with energy. These interactions occur at the atomic level. Atoms of one element react with atoms of another element to form chemical bonds. Atoms absorb and release energy in various forms, including light, heat, or electric current. Every change that you see, hear, taste, or feel involves an interaction of atoms with matter or energy.

How do fireflies make flashes of light?

Light is a form of energy that can be produced by conversion of other forms of energy. We usually associate light with heat, electric current, or both. However, there are some lights that flash without either of these energy sources. How do fireflies produce their bright flashes of yellow-green light on a summer evening?

The process that fireflies use to make light is called bioluminescence. This process, which also occurs in many other organisms, especially in the deep sea, converts chemical energy to light. The chemical energy comes from molecules of adenosine triphosphate (ATP), a chemical that living cells use to store and transfer energy.

De **Definition**

An **enzyme** is a protein that speeds up reactions in living cells. Enzymes take part in the chemical reaction but are not changed by the reaction.

Fireflies have specialized cells that contain a chemical compound call luciferin and an *enzyme* called luciferase. Luciferase is a protein that speeds up the chemical reaction. The flash of light is controlled by the amount of oxygen available to the cell.

The reaction of luciferin with ATP and followed by reaction with oxygen makes a compound called oxyluciferin. Unlike most chemical compounds, the oxyluciferin has electrons that are in an excited state when the compound is formed. As the electrons move to a more stable energy state, they release energy in the form of light. When oxygen is available inside the cells, the firefly's light organs glow yellow or green. When there is no oxygen, the light goes out.

Ff **Fast Facts**

The cool glowing plastic sticks that children (and adults) love to play with use the same principle as firefly flashes. When the stick is bent, a glass cylinder inside is broken, allowing two solutions to mix and produce light from a chemical reaction. However, the chemical reactions used in glowsticks produce light with an efficiency of only about 30 percent compared to an efficiency of 88 percent for the bioluminescence of a firefly.

Living organisms use bioluminescence for several purposes. Fireflies use their flashes to attract mates and also to warn predators that the insect's body contains some really foul-tasting stuff. Some marine animals glow to attract food, and there are a number of animals and bacteria that use bioluminescence for reasons that are not clear.

People have also found ways to use light-emitting chemical reactions based on these natural systems. Biochemists use luciferin as a detector to measure the amount and location of ATP in living cells. The gene that produces luciferase has been implanted in other organisms so that it can be used to follow chemical reactions in cells.

Why don't batteries work well when they are cold?

You use batteries in just about everything—cell phone, iPod, flashlight, car, toys. Batteries provide a way to store energy until you need it and avoid the inconvenience of needing a place to plug everything in. But when it gets really cold, you may find that your batteries let you down, at least until they get warmer. Why would batteries quit working at low temperatures but come back to life when they get warmer?

First, let's take a look at what happens inside a battery. A battery is filled with chemicals that undergo reactions known as *electrochemical reactions.* Each electrochemical reaction has two parts: oxidation and reduction. During oxidation, the atoms of an element or molecule lose electrons; during reduction, the atoms or molecules gain electrons.

If the atoms are mixed together, the electrochemical reaction occurs spontaneously and energy is released, generally as heat. However, inside a battery, the two parts of the reaction are separated from one another. Nothing happens unless electrons have a path to follow between the two sections. A closed circuit provides a path that

De **Definition**

An **electrochemical reaction** is a chemical reaction in which one atom or molecule loses electrons and another atom or electron gains electrons. Electrochemical reactions can be divided into two half-reactions: reduction, during which electrons are gained; and oxidation, during which electrons are lost. Electrochemical reactions are sometimes called oxidation-reduction reactions or redox reactions.

electrons can follow. Electrons flow from one terminal of the battery, produced by the oxidation reaction. They flow through the wire to the opposite terminal, allowing the reduction to take place. In effect, the battery stores chemical energy and then releases that energy as a flow of electrons.

If the terminals are connected by a wire, the electrons flow quickly and the reaction occurs rapidly. The battery and the wire become hot and the chemicals in the battery are consumed quickly. However, if a light bulb or the motor of a DVD player are connected to the circuit, some of the energy of the moving electrons is converted to light or motion. The electron flow is slowed as the energy is removed and the reactions in the battery occur at a slower rate.

Us   **Uncommon Sense**

It may appear that the current flow occurs only between terminals of the battery. However, it is necessary for charges to flow inside the battery as well. If this did not happen, a large negative charge would build at one side of the battery and a large positive charge would build at the other. The battery contains a solution or paste that allows ions to move, making a complete circuit, but which restricts the flow of electrons. Positively charged ions flow through the battery in the direction opposite the flow of electrons. For that reason, if you place a wire with no load across the battery terminals, the battery and the wire both get hot.

On a very cold day, the battery in a car may not have enough current to start the engine. This happens because chemical reactions proceed at a slower rate at lower temperatures. Starting an engine requires a lot of power in a short time. When the chemical reactions proceed too slowly to provide enough power, you get a weak turnover of the engine and it does not start. In this case, the battery operates normally as soon as it is warm because all of the chemical energy of the battery is still available. In very cold climates, battery warmers that plug into an outlet keep the battery warm and make it easier to start the car.

Why do bronze statues turn green over time?

Many cities and towns have parks that include a large bronze statue that has been standing for decades. Its color is generally some shade of green or brown. However, if you buy a new bronze statuette or vase, it usually has a shiny copper or gold-toned surface. Why is the appearance of the metal of the statues so different from that of a new bronze ornament?

Bronze and brass are *alloys* that contain copper along with other metals such as tin or zinc. When copper weathers, it reacts with oxygen and with other materials in the air to form compounds that have properties very different from those of the metal itself.

Most people know that the Statue of Liberty is green. The statue is actually made of shaped sheets of copper attached to a framework. In its pure form, copper is a shiny, reddish-gold metal. However, compounds of copper with other elements or combinations of elements tend to be strongly colored in shades of green, blue, or sometimes pink. When exposed to most air, copper atoms react with oxygen to form two kinds of oxide, one of which is pink and the other black. These oxides react with sulfur dioxide in the air. Copper sulfate is green and copper carbonates are green or blue, depending on the exact composition of the compound. Near the sea, copper chloride may be the predominate salt on the surface. It is copper chloride that gives the Statue of Liberty its color.

De | **Definition**

An **alloy** is a mixture of two elements that has metallic properties that are different from those of either element by itself. An alloy contains at least one element that is a metal and it may contain many different elements. Examples of alloys include brass and bronze, alloys of copper; steel, an alloy of iron; pewter, an alloy of tin; and sterling silver.

Ff | **Fast Facts**

Unlike those made of iron and steel, tools made of copper alloys, such as bronze, do not make sparks when they strike other metal objects or concrete. Industrial facilities in which explosive mixtures can form in the air require the use of bronze tools.

The coating on the metal is called its patina. The patina is a surface coating that tends to protect the metal beneath it from further corrosion.

Although the Statue of Liberty has been standing for over 100 years, attacked by wind, rain, and sea salts, its patina extends only about one two-hundredth of an inch into the copper. If the layer of copper salts were to be removed, the copper beneath would have the shiny appearance of pure metal. However, within months, it would again be green because the surface reaction occurs fairly quickly. Ancient Greek and Roman bronze coins have been found covered with a patina that has protected the surface beneath them well enough to recover the images when the coins were cleaned.

How can a gecko walk across a ceiling?

Geckos are small lizards, recently showing up on television selling insurance, with an amazing ability to climb. A gecko can scurry up a wall, cross a glass window, and even walk along the ceiling. Try doing any of those things and you risk a nasty tumble, if you can even get far enough to fall. Do geckos have some special knowledge of climbing, or do they have tools that you lack?

Us Uncommon Sense

Watching a gecko climb a wall, it is easy to assume that its feet would feel sticky. In fact, a gecko toe feels soft and smooth when you touch it. The ends of the fibers on the foot are made of a protein called keratin, which is similar to the proteins of your hair and fingernails.

People can climb what looks like a solid wall by wedging their hands and feet into tiny crevices. It would seem logical to assume that a gecko does the same thing, perhaps using crevices that are too small to be visible to our eyes. That cannot be the case, however, because the lizards can even clamber up a sheet of polished glass with no significant cracks to wedge into.

Several explanations can be ruled out—suction doesn't work because a gecko's feet don't have a cup shape to push out air; friction might account for the wall but not the ceiling; glue-like sticky stuff would work except the feet don't have any glands to produce it.

To find the answer, scientists have taken a really close look at gecko feet. They found that it's all in the interactions between molecules. The key to being able to stick to the wall doesn't even have that much to do with what gecko toes are made of. Instead it is the shape. Each toe has a network of many millions of tiny hairs, about one millionth of a centimeter long. At the end of each hair are thousands of tiny pads.

The ability to climb, and even hang by a single toe, comes from miniscule electronic interactions, called Van der Waals forces, which occur between molecules. Even molecules that are not polar have small, fluctuating variations in electric charges that occur as electrons move around. These fluctuations cause attractions between molecules. Normally these forces are too small to be noticeable on the scale of objects that we can see. They operate only at very small distances—about the diameter of an atom—so it is hard to get two surfaces close enough to take advantage of the force. The gecko's foot, however, takes advantage of these forces by packing together lots of really tiny pads, each only a hundred or so atoms wide.

The attraction between each pad and the molecules of the surface beneath (or above) the foot is very small. When you have billions of tiny attractions, though, the force adds up. Researchers studying geckos have calculated that, if all the pads were touching the surface at one time, the gecko's feet could hold almost 300 pounds. Because there are so many individual points of contact, the gecko can control the adhesion by varying the amount of contact with the surface. Each contact point only adheres when it is dragged along the surface. That allows it to dash up the wall and not get stuck.

Researchers have used this concept to design a tape that depends only on the shape of millions of tiny structures as an adhesive. They believe that gecko feet may provide the inspiration for new and improved adhesives that hold tight but come off easily.

Ss Science Says

"It is often stated that of all the theories proposed in this century, the silliest is quantum theory. In fact, some say that the only thing that quantum theory has going for it is that it is unquestionably correct."

—Michio Kaku (1947–)

What causes concrete to get hard?

When a truck of concrete arrives at a construction site, its tumbler is filled with a sloppy mixture that is easily poured into a wheelbarrow. The wet concrete is poured into place and smoothed with a trowel. A few hours later, the poured concrete is hard enough to support a person's weight, and rather warm. How does concrete become a hard, stonelike substance?

Concrete is not a recent invention—concrete structures thousands of years old are still standing. It has the strength and durability of stone but can be easily formed into various shapes. Concrete is basically a mixture of cement and water and aggregate—sand, gravel, or other solid particles. The wet mixture can be poured and shaped before a chemical reaction changes its characteristics.

Cement is a mixture of calcium, silicon, and aluminum oxides, often made by heating limestone with clay and then grinding the mixture to make a fine powder. As the concrete hardens, the water reacts with the mineral components of the cement to form crystals. During this process, called hydration, the water becomes part of the crystal structure itself. The water is tightly bound into the cured concrete in a specific ratio to the mineral components, becoming part of the new crystal, and cannot be removed. The interlocking crystals which form around the aggregate convert the wet mixture to a dry, strong, rocklike substance.

It takes some time for the hydration process to completely occur. Depending on the use and amount of strength needed, the concrete must be kept under controlled conditions of temperature and humidity, a process known as curing, for several hours to several weeks. Once it has become solid, water can no longer enter the concrete. If the concrete becomes too dry during curing, it will lose substantial strength.

> **Us** **Uncommon Sense**
>
> Concrete does not harden by "drying." When something dries it loses water by evaporation. In concrete the water becomes part of the final material, so loss of water by evaporation during the hardening process results in weaker concrete. Concrete will even set and cure when it is completely submerged in water.

If you touch concrete during the curing process, you are likely to notice that it is warm. The hydrated cement is a more stable chemical substance than the starting materials. As the chemical reaction proceeds, chemical energy is converted to heat. Removing this heat, or allowing it to dissipate, can be a problem for engineers designing very large concrete structures.

How do heat packs work?

Sitting in the bleachers during a November football game can be uncomfortable if you live in the north. Wouldn't it be nice if you had something warm to keep your hands from freezing? Fortunately, there are heat packs you can buy that generate heat right there where you sit—no electricity, no flames, just warmth. Where does the heat come from?

There are two kinds of heat packs available that depend on changes in chemicals inside the pack. A change the releases heat is called an *exothermic* change. Both types of heat packs use exothermic changes to chemicals. One of them is reversible which allows the pack to be reused. The other involves a chemical reaction that changes the materials in the bag so it cannot be reused.

Reusable heat packs are based on the fact that freezing a liquid material is an exothermic change. This may not be obvious, but if you think about what happens when you make ice cubes, it may be more clear. You place water in the freezer. As it loses energy to the air around it, the water gets colder and the moving molecules move a bit slower. As the water freezes, the moving molecules slow down even more and more energy is released. The solid form of water has less thermal energy than the liquid form, although the temperature remains at 32°F (0°C) during the freezing process, so the process is exothermic. If the freezer did not pump excess heat out into the kitchen, the compartment would become warmer.

De **Definition**

An **exothermic** change is a chemical or physical change that releases energy as heat. An endothermic change is one that only proceeds by absorbing energy from its environment.

If water is very pure, it can be supercooled. If you have very pure water in a very clean glass container, crystals do not form at 32°F. The water can be cooled several degrees below its normal freezing point. However, if you scratch the inside of the glass surface, or even cause shock by tapping the container, crystals will begin to form at the site of the disruption. All of the supercooled water will freeze almost instantaneously.

Reusable heat packs work on this principle. They contain sodium acetate trihydrate, a chemical that supercools very easily, inside a very clean plastic pouch. The freezing point of the liquid is about 130°F (54°C). Inside the pouch is a bent, flexible metal disk. If you flex the disk, the shock causes a few tiny crystals to form. Because the liquid is supercooled, the crystals quickly grow to encompass the whole solution. The heat pad rapidly warms to 130°F as the molecules become more ordered and release heat.

Ff **Fast Facts**

Citrus growers use the energy released by freezing water to protect crops from cold weather. When the temperature drops below 32°F, citrus trees are blanketed with a fine spray of water. As the water freezes, it releases energy and protects the fruit.

Sodium acetate trihydrate heat packs are reusable. If you place the pouch in a pot of boiling water, the crystals melt again to their liquid form. As the pouch full of liquid cools, the liquid again becomes supercooled, ready to freeze again.

The second type of heat pack is filled with a mixture of powdered iron, carbon, water, and salt. To activate the heat pack, an outer pouch is opened that allows air to reach the mixture. A chemical reaction, similar to rusting, begins. The reaction is exothermic, releasing heat over several hours as the iron is converted to iron oxide. Unlike the heat pack based on freezing, this pouch cannot be reactivated and reused.

Why don't matches ignite in the box?

When you think about it, the specifications for a useful match are pretty demanding. It must ignite easily, keep burning long enough to start your fire, and (most importantly) not ignite in your kitchen cabinet until you want the flame. Why do matches light when you strike them but remain stable in the box?

As it happens, the first matches did not remain stable. These matches, invented in the early 1800s, were made using white phosphorus, potassium chlorate, and thickeners to hold it all together. Potassium chlorate is a strong oxidizing agent, which means it readily accepts electrons from other compounds or elements. Phosphorus is easily oxidized. If you mix the two materials, you get an explosive mixture. To light the match, you rubbed it along a rough surface. Friction provided enough heat to cause the volatile mixture to start burning. These early matches tended to ignite at the wrong time and to send sparks flying. They were easy to light, though. When you consider that white phosphorus is extremely toxic and that a single box of matches was likely to contain a lethal amount of it, you realize that these matches were a bit less practical than the matches we use today.

Later, matches used two separate sections on the match head to make a more stable match. Most of the tip was covered with a mixture of potassium chlorate and sulfur in a mixture of clay and glue—very flammable but more stable than white phosphorus. On the tip of the match was a small amount of phosphorus trisulfide, which will ignite when it is rubbed across a rough surface but burns at a fairly low temperature. When this tip ignited, the heat of its flame ignited the hotter-burning mixture behind it. This type of match is still sold, marketed as "strike-anywhere" matches, but they can be hard to find. Strike-anywhere matches can be identified by the two differently colored sections of the match head.

Most modern friction matches are of a type called safety matches. They cannot ignite accidentally because they must be rubbed on a specific surface in order to burn. This surface is generally placed on the outside of the box or on a strip at the bottom of a book of matches. The tip of the match, which is all one color, contains a mixture of antimony trisulfide, potassium chlorate, and binders such as glue. The mixture is flammable but will not ignite just anywhere.

The key to using safety matches lies in the special striking surface. There are two common forms of the element phosphorus—white phosphorus (the highly flammable, very poisonous form) and red phosphorus (less flammable, less toxic). Heat can convert red phosphorus to white phosphorus, which, you may remember, is not very stable in the presence of potassium chlorate. The striking surface has a mixture of red phosphorus and an abrasive material such as powdered glass.

Ss Science Says

"It is the object and chief business of chemistry to skillfully separate substances into their constituents, to discover their properties, and to compound them in different ways."

—Carl Wilhelm Scheele (1742–1786)

When the match is struck, a very tiny amount of red phosphorus is converted, by the heat of friction, to white phosphorus. This little bit of white phosphorus mixes with the chemicals on the match head. The chemical reaction that results ignites the match head, which then ignites the wooden match.

How can a grain silo explode?

You would think that working with flour—finely ground wheat—would be a safe job. The material is nontoxic and is a basic component of our food supply. How could it be hazardous? Well, there is a definite hazard for workers who handle large quantities of flour, or any other powdered grain. There is a risk of a devastating explosion. People have died in explosions that leveled grain mills, grain elevators, and silos. How can something as common and apparently harmless as flour be an explosion hazard?

Flour is made mostly of starch, which is a compound of carbon, hydrogen, and oxygen, and it would be expected to be about as flammable as wood. Although wood burns rapidly under the right conditions, no one expects it to explode. However, the danger of flour in a grain silo is not based on its flammability as much as the size of the particles of flour.

Combustion of a substance in air is a chemical reaction, which occurs when molecules of the substance—in this case flour—combine with oxygen to make new chemical substances—in this case again, carbon dioxide and water. Combustion is a highly exothermic process. That's why a fire in the fireplace warms your living room. The rate of the combustion reaction depends on several factors, including the presence of oxygen and a material that can react with it, a source of ignition such as a spark or a flame, and the ability of the oxygen and the fuel molecules to come together so that they can react. That's where the particle size comes in. The only place that oxygen can react with molecules of the fuel is at the surface of the particles of fuel.

If you're burning a log, or a bag of flour, the surface area of the fuel is very small compared to its total volume. That means the reaction with oxygen is fairly slow. Consider the surface area of billions of small particles of flour dispersed in a large volume of oxygen-rich air. Now you have a lot of places where the two components of the reaction can come together. If a spark or heat source provides a source of energy that causes one of the grains to burn, you have a recipe for disaster. The tiny particle burns almost instantaneously, releasing a lot of heat energy. This energy causes nearby particles to ignite, beginning a chain reaction of burning flour grains. The products of the reaction are gases, which take up much more space than the flour. Hot gases expand rapidly in an explosion. If they are contained in a grain silo or elevator, the rapid expansion can level the structure within seconds of the first ignition.

Ff **Fast Facts**

Flour is not the only material whose dust can form an explosive mixture. Sugar, sawdust, and coal dust have all been involved in deadly explosions. If the particles are fine enough (less than 0.1 millimeter diameter), even metals such as aluminum and iron dusts can explode when they are suspended in air.

How does sunscreen work?

Anyone who has experienced a severe sunburn does not want to repeat the experience. Why does sunlight cause your skin to burn and how does sunscreen prevent sunburn?

The damage to skin is caused by ultraviolet light from the sun. Remember that the radiation from the sun extends far beyond the familiar visible spectrum. Wavelengths that are longer, infrared radiation, can be absorbed by molecules in your body and felt as heat. Wavelengths that are shorter than violet light, ultraviolet (UV) radiation, can pass through the outer layers of your skin. Although UV radiation is very energetic, you don't feel it when it strikes your skin. That's why you can get a sunburn without realizing it on a cloudy day. If the atmosphere absorbs or reflects the infrared light, you don't feel like radiation is reaching you.

Us Uncommon Sense

Many people think that sunscreen is only necessary on hot, sunny days. While clouds absorb and reflect visible light and infrared radiation, ultraviolet radiation passes through them and reaches you even on a cloudy day. Temperature is also not a good indicator of UV radiation. In fact, the danger of exposure increases when there is snow cover because ultraviolet radiation reflects from snow rather than being absorbed by the ground.

However, even though you don't feel the UV radiation, it is still able to do damage. If your skin is dark, you may avoid the burn. The pigment that causes skin to appear dark, melanin, absorbs the radiation near the surface, before it penetrates to the cells beneath. If the UV rays are not absorbed by melanin, then they are absorbed by molecules inside the skin cells. The energy of the absorbed radiation changes some of the molecules and damages the skin several layers deep.

Why is it that the pain and redness of a sunburn do not appear for several hours after the exposure? The symptoms of sunburn are not caused by the damage itself. The pain that you feel comes from your body's efforts to repair the damage. The heat and pain are symptoms of inflammation. The redness comes from enlarging the capillary blood vessels just under the skin so that the body can rush repair cells and materials to the site. The greatest danger of sunburn is the risk of irreparable damage to cells, leading to malignancy.

Sunscreens and sun blocks use chemicals to prevent the ultraviolet radiation from reaching your cells. There are two ways to do this: reflection and absorption.

Some ingredients, such as zinc oxide and titanium oxide, reflect the radiation that strikes them. These compounds are ground into a fine powder and suspended in lotion or cream. They have a white color because they reflect visible light as well as UV radiation. Because it does not allow any radiation to pass to the skin, a thick layer of one of these white creams provides excellent protection from sun damage.

Other compounds used in sunscreens and sun blocks absorb the radiation rather than reflecting it. These ingredients allow visible light to pass through so they are invisible on the skin. However, when ultraviolet light strikes a molecule, the radiation is absorbed, increasing the energy of the molecule. This energy is converted to heat, which does not damage the skin.

Ff   Fast Facts

A glass window can protect you from sunburn. Glass is transparent to visible light, but it absorbs ultraviolet radiation. You can get a sunburn working in the garden, but not the greenhouse—unless the greenhouse has plastic windows.

Like visible light, ultraviolet radiation consists of not just one wavelength, but an entire spectrum of wavelengths. Many of the compounds used in sunscreens absorb only part of the spectrum, so an effective sunscreen may need more than one component to provide complete protection. The labels for these products show that they block both UVA and UVB radiation, which means that they work against the entire range of harmful UV wavelengths.

Part 3 Biological Sciences

Every organism interacts with others around it, forming complex and ever-changing patterns. Growing and reproducing take a lot of energy and matter. Biologists study the constant struggle to be the organism that gets that energy and matter. Ultimately, this struggle leads to the diversity of life that exists around us. And it is all recorded in the patterns of a large molecule we call DNA.

Understanding the natural rules of biology can bring great benefits to humans. Medical scientists have learned how to fight many of the diseases that have afflicted humans for as long as we have existed. We will take a look at some of the questions they ask and the occasionally surprising answers.

Chapter 9

Biology–Plants, Animals, and Others

"It has taken biologists some 230 years to identify and describe three quarters of a million insects; if there are indeed at least thirty million, as Erwin (Terry Erwin, the Smithsonian Institute) estimates, then, working as they have in the past, insect taxonomists have ten thousand years of employment ahead of them. Ghilean Prance, director of the Botanical Gardens in Kew, estimates that a complete list of plants in the Americas would occupy taxonomists for four centuries, again working at historical rates."

—Richard Leakey (1944–)

What do a palm tree, a dolphin, a human, and a mushroom all have in common? Despite the vast differences among these four things, most elementary school students would have no trouble answering the question. All of them are alive.

Biology is the study of living things and the various aspects of life. Biological scientists address many different questions: how do living organisms function; how are they related to one another; how do they depend on one another; how do they survive? These are just a few of the topics of biology.

Beginning in the eighteenth and nineteenth centuries, biologists tried to answer these questions by observing the plants and animals around them. The discovery of microorganisms expanded the field of biology far beyond the organisms observable in our daily lives. Today, the study of life extends down to the interactions of the molecules from which all living things are built. These interactions, occurring constantly inside every living thing, are intimately tied to the big questions that biologists ask.

Are bacteria plants or animals?

Just about anywhere you look, you can find living things. Although there are major differences in the number of organisms you can find in a forest compared to a desert, you can almost always find something alive. You can also classify most of them easily as either plants or animals. But what about the living organisms that you know are there but can't see? Are bacteria plants or are they animals?

De | Definition

Taxonomy is the science of classification. In biology, living organisms are classified by relationships between species. Early taxonomists used observable characteristics of organisms to arrange living things by their similarities. Modern taxonomic classifications are based on the similarities in the genetic material that define each organism.

Swedish botanist Carolus Linnaeus developed the first *taxonomy*, a scientific classification of nature into related categories, in the eighteenth century. He divided the world into three domains: animal, vegetable, and mineral. These classifications remained more or less intact until the middle of the twentieth century. On a large scale, it is usually pretty easy to tell whether something is a plant or an animal. A plant stays in one place and makes its own food; an animal moves around and eats something else. There is the occasional plant that seems to move or animal that seems stuck in place, but it generally works. If you fudge it a bit, you can classify fungi, which don't make their own food, as plants and everything works out.

On the microscopic scale, though, things don't fit so cleanly.

There are the algae, which make their own food, but on the cellular level differ more from plants and animals than they differ from one another. Then there are single-celled organisms that make food by photosynthesis but also move around and eat other organisms.

Bacteria are organisms that have a single cell. With a good microscope, you can find them anywhere from the deepest ocean to the driest desert, and right in your body. In fact, bacteria are so involved with the function of your body that you need them in order to survive. Although they vary in size, you would need to line up about 40,000 of them to make a 1-inch column.

Ff **Fast Facts**

The human body is filled with bacteria. Recent research indicates that your body may hold 10 times as many bacterial cells as human cells. Although most people associate bacteria with disease, many of the bacteria inside our bodies perform functions that are essential to our survival. For example, bacteria play a large role in the digestion of food inside human intestines.

The answer to the original question is that bacteria are neither plants nor animals. Every plant or animal cell has a nucleus, where it manufactures genetic material, DNA. Bacteria, on the other hand, do not have a nucleus. Their DNA floats around inside the cell.

As scientists learned more about microscopic organisms, it became clear that things could not be neatly placed into the plant and animal boxes. A new classification system was adopted in the mid-twentieth century. In that system, all living things were divided into five kingdoms based on the structures of their cells:

- ◆ Animalia—animals

- ◆ Plantae—plants

- ◆ Monera—bacteria and blue-green algae

- ◆ Protista—protozoans and some algae

- ◆ Fungi—fungi

Ss **Science Says**

"Microorganisms will give you anything you want if you know how to ask them."

—Kin-ichiro Sakaguchi (1897–1994)

While this system is useful, it still classifies very different organisms into broad groups. Other classification systems have been proposed, again based on cellular structure.

As more is learned about an even smaller structure, DNA itself, some scientists have proposed that these classifications do not work at all. They have found some relationships across kingdoms that are much closer than relationships within kingdoms. An entirely new classification system is likely to grow as we learn more about life. But in any case, bacteria are not plants and they are not animals.

Are viruses alive?

Anyone who has had the flu knows that his or her body has been attacked by something. The virus that causes flu symptoms not only gets under your skin, it gets right into your cells and takes over. You can't treat the disease with antibiotics, though, as you can a bacterial infection. Antibiotics kill bacteria but they have no effect at all on a virus. Are viruses living things like bacteria?

Believe it or not, this is a question that does not have a clear answer. Viruses are particles, consisting of a DNA or RNA molecule surrounded by a protective protein coat. Viruses reproduce by transferring their DNA to a living cell. Essentially, they hijack the cell, turning it into a factory to produce new viruses. The host cell uses all its energy producing hundreds or thousands of new viruses and then bursts apart, sending them out into the world.

Ff **Fast Facts**

Viruses attack the cells only of specific types of hosts. Although they can attack more than one species—for example, the flu virus can attack both humans and birds—different species may be affected very differently. Also, the viruses that invade plant cells are not able to affect animals. Many viruses prey on bacteria. None of these viruses can have any effect on plant or animal cells. Because of the similarities of DNA in the virus and the host, some biologists propose that viruses are more closely related to their hosts than to other viruses.

Determining whether viruses are alive or not depends on first defining life. It turns out that this is not as simple as it may seem. Since the middle of the nineteenth century, the cell theory has defined life on the basis of cells. Cells are surrounded by a membrane in which the functions of life—such as production of energy, growth, and reproduction—take place. Because viruses do not have cell membranes, do not grow, and do not produce or digest food, viruses are clearly not living things under this definition.

On the other hand, some biologists define life as the ability to pass genetic information from generation to generation. In this sense, viruses are definitely alive. And, although they do not grow once they have been manufactured by a cell, viruses do grow as the cell builds them and they use the mechanism of the cell to convert food into new viruses.

While scientists, and humans in general, have a need to classify things to better understand them, viruses do not fit neatly into either class—living or nonliving.

 Uncommon Sense

Because they can both cause disease, many people think viruses and bacteria are similar and use the term "bug" to describe both. In fact, viruses differ more from bacteria than bacteria differ from humans. Even the smallest bacteria are thousands of times larger than a virus.

Why don't evergreens lose their needles in winter?

In temperate areas, many trees enter fall with a flashy show of color and then lose all their leaves. Others, such as pine, spruce, and fir trees stay green year-round. Why don't these trees lose their needles in the same way that oaks and maples drop their leaves?

The key to answering this question is an understanding of how growing plants take advantage of solar energy. Leaves absorb radiation as light and use this energy to produce sugars that fuel growth and reproduction, and all the functions that support these processes. Near the equator, in tropical areas, the amount of sunlight does not vary much during the year. Plants tend to have large leaves in order to absorb as much light as possible and they tend to grow rapidly, staying green year-round. Leaves tend to stay on the plant until they become so shaded by the leaves above them that they do not produce sugars well.

In temperate regions, the amount of available light varies significantly throughout the year. Deciduous trees—maples, for example—are covered leaves whose shapes resemble those of tropical plants. They collect energy all summer and the plant grows rapidly. In winter, though, sunlight is not strong enough to power these sugar factories.

Then the large leaves can be a problem. Their surface area allows them to lose moisture which may be difficult to replace from frozen soil. Their broad shapes allow them to collect snow and ice whose weight can snap even large branches and trunks. Plants that drop their leaves have a better chance of surviving the cold winds and heavy ice of winter.

But what about the colder regions closer to the poles? Here, the difference between summer and winter is even greater. During summer, days are long and sunlight can last for many more hours each day than in the temperate zones. In the winter, there is very little light available. The trees, however, do not lose their leaves. Why not?

Ff Fast Facts

By any measure, the largest living things are cone-bearing evergreen trees, known as conifers. The tallest known tree is the redwood; the tree with the greatest diameter and the greatest total bulk is the sequoia. Both of these species are native to the western United States.

It takes a lot of energy to grow leaves every year. Because the season is shorter, it is hard for deciduous trees to start fresh each year and still absorb enough energy during the short growing season. Trees that are ready to grow as soon as spring begins have an advantage here. That's why the far northern forests of Europe, Asia, and North America are home to evergreen trees.

These trees are not like the tropical evergreens, however. Instead of broad leaves, these trees, such as pines and firs, have small needles. The needles have waxy coats that retain moisture and their small, rounded shapes do not hold snow the way broad leaves do. Also, the trees have the classic "Christmas tree" shape that allows them to shed snow easily. The trees that thrive in an area are generally the ones whose energy needs balance what is available. In many temperate areas, deciduous and evergreen trees grow side by side, both competing effectively for resources.

Why are most plants green?

If you surveyed a number of people, asking "What color are plants?", you would generally receive the answer "Green." Even though some plants have leaves that are not green, or combine green with other colors, it is true that most plants are green. Why?

Plants are green because the cells in leaves (and sometimes stems) contain chlorophyll, a large molecule that absorbs certain colors of light. This light provides the energy needed to convert carbon dioxide and water into food for the plant, a process known as *photosynthesis*. Chlorophyll uses red and blue wavelengths of light most efficiently and tends to reflect the green wavelengths, so the plant appears to be green.

Why would chlorophyll absorb green more than other colors? This might be a result of the evolution of plants. The light from the sun is not evenly distributed across the spectrum. Certain wavelengths, especially in the red and blue regions, are stronger than others. It may be that chlorophyll dominates because it absorbs energy most efficiently where it is strongest. Plants that developed other mechanisms for absorbing light may have lost out because chlorophyll was more efficient.

Chlorophyll is not the only compound in leaves that gives them color, but it is the most visible. In temperate regions, we can see some of the other pigments in the fall. As photosynthesis becomes less efficient when light and heat drop off, the plants quit making chlorophyll. Then the reds, yellows, and purples of other pigments in the leaves make their mark.

> **De** | **Definition**
>
> **Photosynthesis** is the process used by green plants, as well as some bacteria and algae, to produce glucose from carbon dioxide and water. The term *photosynthesis*, literally "building with light," is used because the energy for the chemical reaction comes from sunlight.

How do chameleons change color?

Chameleons are African lizards that can change color, an ability that is unusual among animals. Each chameleon can display a range of colors—brown, black, blue, green, red, yellow, or white. How can a chameleon change colors so easily while most animals can be identified by their constant color?

Its ability to change color lies where you would expect—right in its skin. A chameleon's skin has several layers. The protective outer layer is transparent and colorless. Just below it lies another layer of skin, made of cells called chromatophores, which contain

> **Us** | **Uncommon Sense**
>
> A chameleon does not change its color to match the background. In fact, many of the shades that a chameleon takes when it is not agitated or excited provide camouflage in its natural environment. Generally, chameleons going about their daily business undisturbed will be various shades of green, blue-green, or brown.

red and yellow pigments. The third layer contains chromatophores rich in melanin, the same pigment that accounts for the different shades of human skin. The melanin layer creates black and brown colors and it also reflects blue light. Delving even deeper, you find the fourth, and final, layer of skin. This layer reflects the light that reaches it.

The color of the chameleon's skin changes as the chromatophores expand and contract. The temperature and the brightness of light can affect the size of the chromatophores and, with them, the color of the animal. Most of the control, however, comes from nerve impulses from the brain and changes in hormone levels in the cells.

When a chameleon is cold, the red and yellow cells shrink and the brown/black cells expand, giving the skin a dark color which absorbs more energy from light. When it is defending its territory or interested in mating, it produces bright, showy colors. The meaning of a particular color or pattern of colors varies with the species of chameleon.

What is the oldest living thing on Earth?

The oldest person whose age could be confirmed lived 122 years. There are a few animals that live longer. The oldest giant tortoise whose age has been confirmed lived 188 years. You won't find the oldest living things among the animals, though. What is the oldest living thing?

The age of a tree can be determined by counting its growth rings. Many large trees are very old. Counts of the ring growth in core samples have shown that some redwoods and sequoias have been growing for more than 2,000 years. Until recently, the oldest known organism was a bristlecone pine in California, which has been named "Methuselah." A count of its growth rings from a core sample taken in 1957 showed that the tree was 4,723 years old.

But that is not the end of the story. Botanists have found an even older plant in southern California. Creosote bushes do not have the grand size of a sequoia, or the gnarled, ageless appearance of the old bristlecone pine. In fact, an old creosote bush looks like a ring of scruffy shrubs with a distinctive odor.

The development of genetic testing has shown that the shrubs in a ring actually comprise one plant. The original stem of a creosote bush dies after a few decades, leaving several new stems to take its place. Although the stems appear to be separate bushes, they share an interconnected root system. As these stems are replaced and die, the ring moves outward. The plant develops into separate bushes, but each of them has a legitimate claim as being the original plant. All of the bushes are genetically identical. One ring of creosote bushes has been measured by carbon dating to be more than 11,000 years old, twice the age of the oldest known tree.

Even more recently, another discovery threatens to make the creosote bushes rank as toddlers in the old-age contest. Bacteria have been recovered from ancient ice samples far below ground in Siberia. These cells appear to have been alive and functioning at a very slow rate for more than 600,000 years. For now that is the record, but who knows what may be found next?

Ff Fast Facts

Although each ring represents one year of growth, tree rings are not all the same size. During dry years trees have less growth than during moist years, so the ring is narrower. By studying the patterns of ring size, scientists can determine the climate of a region, tracing periods of rain and drought over hundreds of years. Comparisons of the patterns can even be used to date logs of trees that died or were cut many centuries ago.

How can grasshoppers jump so far?

Picture an athlete in the Olympics, preparing to make a world-record long jump. He takes a long run, getting to sprint speed, and launches at the line. A really good jump carries him almost 9 meters, about four-and-a-half times his height.

Now watch a grasshopper jump. From a standing start, the 4-centimeter-long insect springs forward, landing more than a meter away, about 40 times its body length. And that's an average jump, not an Olympic record. Why can a grasshopper jump so much farther than a human compared to its size?

There are several reasons why an insect can jump proportionally farther than a person. The two organisms have different body designs and different methods for getting oxygen to muscles. The most important factor, though, has to do with changes in scale as size increases. The force that a muscle can exert increases as the cross section of the muscle increases. The cross section is an area measurement, a square of length. If the grasshopper doubled in length, keeping the same body proportions, its muscle cross section would increase by four. If it tripled in length, its cross section would increase by nine, and so on. A 6-foot grasshopper would have about 2,500 times as much jumping power in its huge rear legs.

But here is where the scaling problem comes in. Area is squared as length increases, but volume is cubed. As the length of the grasshopper increased by a factor of 50, its volume increases by a factor of 125,000. That means its mass also increases by that factor. Although the muscles are stronger, they must propel 50 times as much mass per cross-sectional area. A grasshopper the size of an Olympic jumper might not even be able to jump as far as the human.

Why are moths attracted to light?

If you sit outside on a dark night with the porch light turned on, you are almost certain to see many moths fluttering around the light. It seems strange that a nocturnal insect would want to be close to a source of light because there is not any natural source of nighttime light that would be useful to moths. So why are the moths attracted to light?

This question is an example of an observation that can be misleading. Because many moths gather around the light, we assume that they are attracted to the light. There is no evidence that the moths approach light because they are attracted to it. In fact, many species of moths are repelled by light and fly away from it.

There are several theories about the interaction between moths and light. Most scientists who study moths think that the moths approach the light as a result of confusion, not attraction. It may be that moths, which have good eyesight, normally use natural light sources—the moon and the stars—to orient themselves so that they fly in a straight line. Consider the effect on the insect's navigation, if this theory is correct. A moth would fly toward the light, not because it is trying to approach it (a moth cannot approach the moon) but because it sees a reference point. Unlike the natural light sources that the moth evolved to use for reference, the light gets closer and closer, so the reference becomes confusing and the moth flies in circles.

Another possible explanation is that the moths are looking for food. Some moths are attracted to night-blooming flowers. Many of these flowers are large and white, reflecting as much light as possible. If this is the case, it might explain why some types of light seem to be more attractive to moths than others. For example, ultraviolet and blue wavelengths seem to be more attractive than yellow wavelengths. Flowers may not reflect all wavelengths equally.

Yet another explanation could be that the moths come upon the light by accident. Their eyes have many facets that collect light. As the moth approaches the bright light, it is temporarily blinded and its eyes need to adjust, just as your eyes must adjust to a sudden bright illumination of a dark theater. After adjusting to the light, the moth is then unable to see in the dark as it leaves the light and it blunders back to the light.

Uncommon Sense

The ultraviolet light traps that "zap" bugs in many backyards are not very effective at preventing bug bites. Mosquitoes do not show any great tendency to approach them. Most of the time, that satisfying "bzzzzt" that you hear is a moth making its way to the electrical circuit as it flutters around the light.

Does a camel really store water in its hump?

It's easy to understand why people think the big hump on a camel's back is a water container. The animal can travel many days in the hot, dry desert without drinking any water and then gulp down 20 gallons all at once. Is this water actually stored in the hump?

Although the camel can store a lot of water in its body, the hump is not a stock of water. The camel's hump is a large fat deposit—a store of food rather than water. The fat in the hump allows the camel to go for long periods without eating. When food is short, the fat is metabolized for energy, in the same way that a store of fat carries a bear through a winter without eating. Camels can go for up to two weeks without eating by relying on this reserve. It is uncertain why the camel stores fat on its back, but one possibility is that the fat also acts as insulation, protecting the animal from the hot sun.

Even though water is not stored in the hump, a camel can tolerate long periods without water. For most mammals, drinking the equivalent of the camel's 20-gallon chug would dilute body fluids, perhaps fatally. The camel, however, can adjust to the swing in water concentration in its body. In addition, its body temperature can range from 94°F to 105°F. This allows the camel to adapt to the heat of the day and not start sweating until long after most mammals would have needed to expend moisture to cool their bodies.

Ff **Fast Facts**

A camel is perfectly adapted to the desert. It has three eyelids to protect its eyes from the sand and dust of a desert sandstorm and it can close its nose to protect its breathing passages. A camel's broad, leathery feet can find traction in rocky soil and soft sand.

Why are fossil shark teeth common but not shark skeletons?

We know that even the great white shark would be dwarfed by sharks of the past, especially the giant megalodon, which is estimated to have grown to more than 50 feet in length. Size estimates for this giant fish are based on fossil teeth. No one has ever found a fossil megalodon skeleton, even though dinosaur skeletons many times older are common. Why is it that there are no fossils of ancient shark skeletons but fossilized shark teeth are very common?

Sharks are amazingly agile fish, able to make a tighter turn than other fish the same size. One reason for the shark's maneuverability is its skeleton. Unlike most other vertebrates, the shark does not have a skeleton made of bone. Instead, it is made of cartilage, the strong, lightweight material that forms the shape of your nose.

Cartilage is much lighter than bone, which helps keep the shark from sinking and is more flexible for turning efficiency. In addition, the shark's swimming muscles are not connected to the skeleton, but instead they connect directly to its tough skin, increasing the efficiency of the muscles. However, because the cartilage is not hard like bone, it does not tend to form fossils. Very few fossil shark skeletons have been discovered.

Us **Uncommon Sense**

The belief that sharks do not get cancer has led to a market for shark cartilage as a cancer treatment. In fact, sharks do get cancer, although their immune systems seem to be particularly good cancer fighters. Scientists are doing research to determine whether sharks produce compounds that would be effective in treating human cancers. However, there has never been a scientific study that showed that ground-up shark cartilage has any effect as a treatment.

The shark's teeth are hard, like your own, and easily form fossils. In addition, sharks have lots of teeth and they make new teeth throughout their lives. Sharks do not have one row of teeth like humans. Instead, they have several rows of teeth attached to the jaw along with 5 to 15 rows of spares. As teeth become worn or damaged, they fall out. When a tooth is lost, a new one takes its place, generally within a day. A single shark can produce more than a thousand teeth every year. That makes a lot of material for fossils.

Although the shape and size of sharks' teeth vary from one species to another and even from one part of the jaw to another, the size of the teeth can be used as a measure of the size of the fish. We know that the megalodon was a truly huge shark because it had huge teeth. While the teeth from a great white shark may measure about 2½ inches long, some megalodon teeth exceed 7 inches.

Why do animals migrate?

An aerial photograph of a massive migration, whether it is caribou in North America or wildebeest in Africa, is an impressive sight. Thousands of animals move together as one large herd. Why do animals migrate and how do they know where to go?

Some animals migrate extremely great distances, traveling thousands of miles annually. Migrations consume large amounts of energy, so they are generally undertaken for important reasons: reproduction, food supplies, or warmer climates. Antelope travel across the plains, looking for greener pastures. Geese fly from the abundant summer food supplies of the northern tundra as days shorten and the temperature drops, only to return in the spring when the days lengthen. Salmon fight the currents of swift-flowing streams to return to the place of their birth, laying the eggs of the next generation.

Us **Uncommon Sense**

Migrating lemmings do not commit mass suicide by flinging themselves off cliffs into the sea. Lemmings do migrate from one area to another when their population exceeds the food supply. During the migration, they swim across rivers and some of them drown. This may have been the origin of the suicide myth. The most famous migration of lemmings over a cliff occurred in a 1958 movie. The film was made far from the sea and the "suicidal" lemmings were actually tossed over a ledge by the movie's producers.

Although it is not too difficult to figure out why a particular migration takes place, the how can be a bit tougher. How can a monarch butterfly travel from Canada to Mexico, or a Pacific trout swim a thousand miles to the mouth of the stream in which it was spawned?

Ff **Fast Facts**

The phenomenon of migration is not limited to animals. Human migrations have led to populating the entire planet. Like animal migrations, human migrations occur when people seek escape from threatening conditions, such as war, or search for places with better resources, such as the migration across North America in the nineteenth and twentieth centuries. There are even some seasonal human migrations. For example, migrant farm workers follow the harvest seasons in many countries.

Researchers have studied the physical characteristics of migratory animals and the characteristics of their movement to come to conclusions about how animals navigate. Many migrating birds, including ducks, use the skies to orient themselves, observing the positions of the sun or the stars. Some animals, including loggerhead turtles, have sensory organs that detect Earth's magnetic field, a built-in compass. Salmon detect their home stream after traveling vast distances in the ocean by its smell. It is likely that many migratory animals use several environmental cues, including these to find their way around. Only humans use printed maps, but the animals don't seem to miss them.

Chapter 10

Biology–Humans

"Biology will relate every human gene to the genes of other animals and bacteria, to this great chain of being."

—*Walter Gilbert (1932–)*

The biology of humans is related to that of other animals, but, of course, there are significant differences. Studies of anatomy and physiology have led to a breakdown of the body into interacting systems, each with its own role in making the body and mind function. Some of the systems—circulatory, respiratory, and, of course, reproductive—are quite familiar (but not always completely understood). Others may be less known but are certainly no less important:

◆ The nervous system, the body's control center, consists of the brain and the network that connects it to senses and muscles.

◆ The gastrointestinal system manages the conversion of food into materials useful to the body and eliminates wastes.

◆ The respiratory system brings in oxygen and eliminates carbon dioxide.

◆ The circulatory system works with the other systems to carry nutrients and oxygen to cells and carries wastes away.

◆ The musculoskeletal system is the framework of bone and muscle that gives structure and motion to the body.

◆ The immune system recognizes and destroys foreign cells and tissues that may be harmful.

◆ The endocrine system produces the hormones that carry signals between systems and regulate many of the body's functions.

◆ The integumentary system, including the skin, hair, and nails, covers the body and interacts with the outside.

Why are there different types of blood?

During a crisis on a television hospital show, patients stream into the emergency room. Not knowing the patient's blood type, the doctor calls for O-negative. In most ways, everyone's blood is the same. It consists of two main types of cells—red and white—suspended in a clear salty liquid—plasma. On a microscopic level, however, the red blood cells are not all alike. If the wrong types are mixed together, they form clumps of cells that can interfere with blood flow, with possibly fatal results. Why do people have different types of blood?

De **Definition**

An **antigen** (*antibody generator*) is a compound that stimulates a defensive response from the immune system. Most antigens are either proteins, which are chains of amino acid molecules, or polysaccharides, which are chains of sugar molecules. Many antigens are parts of bacteria or viruses.

The difference in blood types is a result of chains of different sugar molecules that are attached to the surface of red blood cells. These attachments, known as *antigens*, identify the blood cell to the immune system.

Two distinct types of antigens exist in human blood, labeled A and B. If a red blood cell has the A antigen, it is labeled Type A. There are four possible blood types: Type A, Type B, Type AB (both antigens), and Type O (neither antigen). If blood containing one of the antigens is transfused into a person whose cells do not have that antigen, the immune system reacts as if the new cells were an invading infection. As the body attempts to destroy the invaders, the blood forms dangerous clumps.

After the development of the ABO system of classification, a third antigen, called the Rh factor (because it was first observed in the blood of rhesus monkeys) was discovered. Red cells with the antigen are called Rh positive and those without it, Rh negative. Combining this classification with the ABO system gives eight distinct blood types. Because Type O negative blood has none of the antigens, it can be given to a person of any blood type without risk of an immune response, so people with O-negative blood are called universal donors. If you have Type AB-positive blood, your immune system does not respond to any of the antigens, and you are a universal receiver.

Blood type is a genetic trait, inherited from your parents. A parent with Type A blood can have children with Type A or Type B, but not Types AB or O. Although it is not certain why the blood types developed, the frequency of particular types of blood varies among populations originating in different places on Earth. Research indicates that blood type may be linked to susceptibility or vulnerability to certain diseases. For example, people with B or O blood have a slightly lower risk of certain cancers, while people with the A or B antigens (or both) have a lower risk of contracting cholera or plague. The antigens may have developed as part of the body's immune response to diseases.

> **Us** | **Uncommon Sense**
>
> The Blood Type Diet is based on the premise that a healthy diet depends on your blood type and recommends specific food groups based on blood type. However, the proponents of the diet do not provide any evidence that clinical trials have been performed to support their claims. Most dieticians and nutritionists consider the diet to be based on pseudoscientific claims.

In 2007, an international research team announced that they had discovered an enzyme that can remove the antigens from red blood cells. Although extensive testing will be necessary before this process becomes acceptable for general use, it may allow any type of blood to become Type O. This would greatly increase the efficiency of blood banks and the safety of transfusions.

Why is it impossible to go without sleep?

You can choose to do many of the things necessary to sustain your body, even though the consequences may not be pleasant. Food is no problem—you can refuse to eat for quite a while and still recover. Refusing water can cause death by dehydration, but it can be done. You can't refuse to breathe because your body will take over and do it for you, even if you pass out first. And you can't refuse to sleep. No matter how hard you try to stay awake, it is certain that you will eventually nod off. Why is it impossible to go without sleep?

Experiments with rats have shown that complete sleep deprivation leads to death even faster than starvation. However, it is almost impossible to prevent sleep after a certain point. Even stopping short of death, though, sleeplessness can wreak havoc on the body. Eventually, you become irritable, then forgetful. It becomes impossible to complete even the simplest task, let alone something complex, like driving a car. Extended periods of inadequate rest lead to reduced immune system response, fluctuating blood pressure, and changes in metabolism.

Interestingly, for something so important to our well-being, we really don't understand sleep that well. There is no obvious chemical change, such as the buildup of carbon dioxide when we don't breathe, to explain the changes. Sleep scientists do know, however, that when our bodies shut down and we sleep, the brain keeps going.

Ss **Science Says**

"I think it's a very valuable thing for a doctor to learn how to do research, to learn how to approach research, something there isn't time to teach them in medical school. They don't really learn how to approach a problem, and yet diagnosis is a problem; and I think that year spent in research is extremely valuable to them."

—Gertrude Elion (1918–1999)

In the 1950s, scientists discovered that there are two distinct brain states during sleep—rapid-eye-movement (REM) sleep and non-REM sleep. REM sleep, as the name implies, is characterized by movement of the eyes beneath their lids. During non-REM sleep, the brain appears to go into a slower state, like an idling engine. Breathing and heartbeat are regular and there are few dreams.

During REM sleep, however, the brain is very active and neurons appear to react very similar to the waking state. This is the part of the cycle during which dreams occur. As we dream, the parts of the brain that control motion in the body operate almost as they do during when awake, but the neurotransmitters that carry signals from the brain to the muscles are inhibited, with the exception of those linked to motion of the eyes.

Sleep research is a complicated process but scientists are starting to develop some hypotheses about our need to sleep. Non-REM sleep is a period of lower metabolic rate and lower brain temperature. This appears to be the period during which the brain can undertake repair work inside its cells and restock some enzymes that it needs in order to function properly. During REM sleep, the cells function normally, but the neurotransmitters are turned off. This may be the time during which the key links in the neurons can be restored. Researchers believe that these links play an important role in controlling mood and learning. The brain may also use its sleep time to organize memories and data from the day and to develop the brain itself as it cuts the rest of the body out of the process.

Ff | Fast Facts

Although it is not clear whether insects and other invertebrates sleep, researchers have found evidence that mammals, reptiles, and birds all need sleep. At least in mammals, the amount of sleep needed depends on body size. An opossum sleeps about 18 hours every day, while an elephant manages quite well with 4 hours or less. Marine mammals continue swimming while they sleep.

Why doesn't it hurt to cut your hair and fingernails?

Hair and fingernails both grow but you don't feel any sensation when you cut them. Are these parts of the body made of cells or something else?

Your hair and your fingernails are made by living cells, but the material itself is not living. Both hair and nails are built from layers of dead cells. The main component of both substances is a fibrous protein, known as keratin, which is also used by many animals to make hooves and horns (unlike antlers, which are made of living bone cells).

Fingernails grow from their base where a living structure, the matrix, produces the nail cells beneath the protective skin of the cuticle. These cells are formed in layers, making the hard plate of the nail. As the nail grows, at a rate of about 3 millimeters per month (half that for toenails), these cells are pushed forward. As they move, the cells die, leaving the layers of hard protein.

Your body produces hair in a similar way. Each hair grows from a follicle that has a cluster of cells that produce hair cells. As in fingernails, the cells die as they move away from the growth point. Because the cells are not living, damaged hair and nails cannot heal or be repaired. Each hair is made of several layers. The keratin fiber is surrounded by the cuticle. The dead cells of the cuticle overlap like roof shingles and protect the protein fibers inside from damage.

Us   Uncommon Sense

There is a belief, fostered by many mystery novels, that hair and fingernails continue to grow after a person dies. This is not the case—the matrix and follicle cells die along with all the others. Fingernails and hair (particularly beard hairs on a shaven face) seem to grow because the skin around them shrinks as it dries. The exposed length increases, although the actual length of the hair or nail does not change.

Why doesn't stomach acid dissolve the stomach itself?

When you eat a meal, the digestion process starts with your saliva as you chew. The heavy-duty work, breaking down the cells that form plants and animals, falls to the stomach. The acid in your stomach is strong enough to reduce a steak dinner to liquefied mush. So why doesn't the acid dissolve the stomach itself?

The answer may be surprising—it does. The inside of the stomach is lined with protective cells called epithelial cells. The most common of the epithelial cells is a type that produces mucus—a thick protein covered on the outside with sugar molecules held to it by chemical bonds. The sugars are able to resist the effects of acid much better than protein, so the mucus layer protects the stomach lining from itself.

Even so, acid constantly works its way to and through the layers of the lining. This is where a second layer of protection comes in—and where the stomach eats itself. The outer layer of epithelial cells is constantly being attacked, but new cells move up from beneath to replace damaged cells. In a healthy human, about 500,000 epithelial cells are destroyed each minute. Over the course of about three days, the entire lining of the stomach is replaced in a constant cycle of cell death and birth.

If the acid manages to breach the mucus layers and the epithelial cells, it can cause dangerous and painful deterioration of the muscles that cause the stomach to work. Ulcers and stomach cancer are two of the most severe consequences of a breach of the protective layers of the epithelial lining.

Ff Fast Facts

Until fairly recently, doctors believed that stomach ulcers are caused by spicy foods and/or stress. In the 1990s, researchers showed that the ulcers are actually the result of a bacterial infection in the stomach lining. The bacterium survives the stomach acid by secreting enzymes that neutralize the acid around it. It then burrows into the mucus of the stomach lining where it is further protected. Ulcers form when the bacteria weaken the lining and allow acid to reach the tissue beyond. Now, the standard treatment for stomach ulcers is an antibiotic.

How many different types of taste can you sense?

Taste is one of our windows into the world around us. Besides providing the ability to enjoy our food, it has a valuable protective function. Many toxic chemical compounds (but certainly not all) have a taste that is universally unappealing or even repulsive. Sometimes taste preferences change with nutritional needs. How does your body register the taste of a food?

The basic sensors for taste are on your tongue and around the other parts of your mouth. The bumps on your tongue contain bundles of taste buds. There are about 10,000 taste buds altogether. Over a two-week period, they will all be replaced as they wear out. Each taste bud consists of dozens to hundreds of taste cells that receive certain types of chemical information and transmit them to the brain through nerve networks that end in the taste buds.

There are a large number of different receptors on the taste buds that bind to specific chemicals, but essentially there are five tastes recognized by human senses:

- ◆ Sweet
- ◆ Sour
- ◆ Salty
- ◆ Bitter
- ◆ Umami

These are all familiar terms, except umami, which is sometimes called savory. It is the characteristic taste of a rich, meaty broth or strong cheese.

These five sensations are only part of the story, though. Think about how tasteless food seems when you have a bad cold with a stuffy nose. Taste is a cooperative sense, working with the sensors in your upper nose that determine smell. In fact, based on the number of sensors, the sense of smell may be more important than the taste buds in the perception of the flavor of something. There are about 100 million sensors at the back of each nasal cavity. Molecules travel upward through openings in the roof of your mouth to these sensors. If you cannot smell your food, your taste sensations are very limited.

Us **Uncommon Sense**

Remember the map of the tongue that showed up in your high school biology book? It showed which parts of the tongue were able to detect sweet tastes, which could detect salty, and so on. If you ever tried it, you may have found that the map did not match your tongue. Guess what? It doesn't match anyone's tongue. It appears that a combination of incomplete research and a poor translation led to the map. Controlled research has shown that all of your taste buds have receptors for all five tastes.

What causes jet lag?

Frequent travelers know the signs: you have traveled across the Atlantic, the sun is coming up, a busy day awaits, and all you want to do is close the curtains and crash. Jet lag has struck again. No matter how many times you travel or how many ways you try to trick your body, you just can't avoid it. What causes jet lag, and is there any way to avoid it?

Jet lag occurs when you cross multiple time zones in a short time. Just as the clock beside your bed operates on a 24-hour cycle (and must be reset when you change time zones), you have a built-in clock that cycles daily. If you cross several time zones quickly, your internal clock doesn't match the natural daily clock, based on the rising and setting of the sun. Your body gets confused, giving you a headache and an upset stomach, and making it difficult to concentrate. In general, jet lag occurs with time zone changes of three hours or more, but it is extremely variable from one person to another.

This internal time sense, called a *circadian rhythm*, is not just a human phenomenon. Animals also function in a day-to-day world and have a time sense that is built into conscious and unconscious schedules. In fact, the subjects of the original studies of the genes that control these biological rhythms were fruit flies. Researchers have even studied fungi that produce chemicals to protect their cells from ultraviolet light just *before* sunrise, even if the fungus is moved indoors.

De **Definition**

A **circadian rhythm**, or circadian cycle, is an approximately 24-hour internally regulated cycle of functions, including sleeping and waking, growth, and hormone production in an organism. Circadian rhythms have been observed in animals, plants, and bacteria. "Circadian" comes from the Latin for "about a day."

In humans, the circadian rhythm is controlled by a tiny part of the brain near the optic nerves. Genes that control our internal clocks produce proteins that break down over time. Time is not the only factor in this breakdown. Changes in exposure to light can reset the clock, so each morning an adjustment takes place, just like someone looking at a watch and setting it against a reference time. Little adjustments each morning keep us in time if our clocks get out of synch. Light striking the retina sends signals to this part of the brain, shutting off the production of melantonin, a hormone that causes people to feel drowsy. Other body systems also respond to signals from the brain, controlling functions such as blood pressure, urine production, and production of other hormones.

Ff **Fast Facts**

Working a rotating shift exerts the same effects on the body as changing time zones. The risks of disruption of circadian cycles also seem to increase with age and they can be severe. Research indicates that women whose jobs cause chronic changes in schedule for longer than 15 years have higher levels of breast and colorectal cancers.

So why does jet lag go away after a few days? It turns out that the adjustments triggered by exposure to light are a bit limited and can only move the clock by one or two hours each day. That means a change of six time zones will take about three days of adjustment. In the meantime, your body acts as if you were a few thousand miles east or west. That is jet lag.

If humans don't need an appendix, why do we have one?

Just above your large intestine is a small worm-shaped organ, about 3 inches long, called the appendix. You seldom hear about the appendix unless it becomes inflamed. Then there is a risk of the organ bursting, spreading bacteria throughout the abdomen and causing severe pain and, sometimes, death. An inflamed appendix can be removed with no apparent effect on the body. If it causes problems, and we can do without it, why do people have an appendix?

Until recently (and possibly even now), most doctors would have told you that the appendix has no useful function. In fact, it is almost a hazard. About 7 percent of Americans have an appendectomy during their lifetime and appear to suffer no ill effects. The presence of an appendix is explained as a "leftover" of evolution, useful in some mammals but not in primates, a group that includes humans.

In some mammals, such as rabbits, the appendix helps digest cellulose, a major component of grass and the stems and leaves of many plants. Specialized bacteria live in the appendix and break down the cellulose into compounds that can be absorbed by the animal's digestive system. Primates, however, do not eat grass. They live on insects, meats, and plant foods that contain starches and sugars instead of cellulose. According to the hypothesis, the appendix was once a useful organ to our far distant (preprimate) ancestors. Now it remains because there is no evolutionary advantage that would cause people without an appendix to be better adapted to their environment. Therefore, there is nothing that would lead to its disappearance.

Recent research, however, suggests that the answer is not that simple and that the appendix does have a use, after all. The immunologists who performed the study point to the number of bacteria in our intestines that are needed for our survival. These bacteria play a major role in breaking down food so that it can be used by our bodies. Researchers have found evidence that the appendix provides a place for these essential bacteria to live. Sometimes diseases that are accompanied by severe diarrhea cause such extreme emptying of the intestines that the populations of bacteria are substantially depleted. These researchers have formed a hypothesis that the appendix then serves as a reservoir of "good" bacteria, releasing them into the intestines and speeding recovery. Because the reservoir is seldom used, and it is possible to recover without it, people can survive without an appendix.

Everyone agrees, though, that an inflamed appendix should be removed because the risks associated with appendicitis are greater than the benefits of having an appendix.

Ff Fast Facts

The appendix is not the only organ that sometimes seems to be more trouble than it is worth. The tonsils at the back of the throat are lymph nodes that provide a first line of protection against infection. However, like the appendix, they are prone to become infected themselves. A generation or two back, removing the tonsils was almost a routine part of childhood, with no obvious decrease in immunity. Pediatricians today tend to recommend a tonsillectomy only when chronic infections interfere with the general health of a patient.

Why do women live longer than men?

Along with medical advances and increased availability of basic needs, the average life span of humans has increased significantly over time. There is a significant variation among different populations but, with only a few exceptions, women tend to live longer than men. For example, in the United States, the average for women is 80.1 years, while for men, it is 74.8 years. How can we explain the correlation of life expectancy to gender?

This question is not easy to answer, mainly because there are many factors to be considered and, in general, they are not easily controlled in an experiment. Part of the answer seems to be tied to societal roles. For example, in most societies, men are much more likely to fight wars, work in physically risky jobs, and generally be more aggressive, leading to death in conflicts. However, there is some evidence that this explanation does not account for all of the difference. Part of the explanation may be biological, which would not be surprising considering the number of physiological differences based on gender.

Additional evidence that biology explains part of the difference comes from studies of animals. Females have a greater average life span among elephants, mice, and even fruit flies. Several explanations have been proposed and researchers are looking for evidence that one or more of them can explain the longevity difference in animals and to determine how they relate to humans.

One difference is fairly obvious. Sex hormones play a large role in the differences between male and female. Estrogen, for example, helps to eliminate cholesterol, while testosterone increases the levels of low-density lipoproteins (the so-called "bad cholesterol"). This could account for part of the difference in mortality due to heart disease. Women also have a slower metabolism, on average, than men. Many studies of animals have related slower metabolism with longer life span.

Another major difference between men and women is genetic. In many animals, including humans, females have two X chromosomes, while males have one X and one Y chromosome (see Chapter 11). A number of diseases—hemophilia, for example—are much more common among men than women because a gene on the second X chromosome blocks the defective gene that causes the disease. The X chromosome also includes a gene that is used in the repair of damaged DNA. Because a male has only one copy of this gene, if it is defective he may experience more of the effects of aging and disease as unrepaired mutations accumulate during his lifetime.

Ff **Fast Facts**

Research shows that reducing food intake to the minimum amount needed for proper nutrition extends the life spans of many animals, including monkeys, rats, mice, and fruit flies. Interestingly, calorie restriction shows no effect on the life span of the housefly. Researchers do not know if humans can increase the length of their life by cutting calories to the bare minimum, although there are some indications that, even in nonobese people, reducing calories improves some indicators of health.

It is important, though, to keep in mind that all of these explanations address the average longevity. For any individual, the potential to live a long and healthy life is increased by a healthy lifestyle.

"The brain is the last and grandest biological frontier, the most complex thing we have yet discovered in our universe. It contains hundreds of billions of cells interlinked through trillions of connections. The brain boggles the mind."

—James D. Watson (1928–)

Biology—Genetics and DNA

> "If we examine the accomplishments of man in his most advanced endeavors, in theory and in practice, we find that the cell has done all this long before him, with greater resourcefulness and much greater efficiency."
>
> —*Albert Claude (1898–1983)*

Have you ever looked at two siblings and known right off that they were related? Their body structure, facial features, colors of the skin, eyes, and hair are almost the same. It is clear that they have the same parents. But they may have a third sibling whose appearance is greatly different. Why would two children with the same parents look almost identical in some cases but extremely different in others?

The first clues to the answer came in the 1800s when Gregor Mendel began studying heredity in pea plants. He found that there are rules to inheritance of characteristics from parent to child, not just in humans but in every organism that reproduces sexually.

Since the time of Mendel, we have learned a lot about why traits are inherited. We know now that the rules explaining what features are passed from one generation to the next are all stored, as if they were a written code, in molecules of DNA. Genetics is the science that studies heredity and variation in organisms and how DNA carries information through one generation after another.

What is the purpose of DNA?

The cells of every living organism, from the simplest bacterium to humans, contain DNA molecules. These are very long strands (the DNA molecules in a human cell would measure about 1 meter long if stretched out) made up of chains of smaller molecules. What is the purpose of these DNA molecules?

Inherited traits are passed from one generation to the next through DNA. It stores the instructions a cell needs in order to make proteins. These proteins are the keys to a wide range of chemical reactions that direct the metabolism, growth, and specialization of living cells.

A molecule of DNA has a backbone structure with a sequence of four different molecules, called bases, attached to it. The sequence of the bases codes information about the organism in the same way that the sequence of letters of the alphabet codes the meaning of this sentence.

Every function of a cell and each trait of the organism is controlled by a *gene*, or sometimes a combination of genes, located on a particular section of DNA. Each gene is one segment of the DNA sequence, described by the pattern of hundreds or thousands of bases.

De Definition

A **gene** is a region of a DNA molecule that controls the production of one or more specific proteins. All of the gene's information is contained in the sequence of bases in its part of the DNA molecule. Each gene makes up one unit of inheritance.

A simple bacterium may have a single strand of DNA holding about 1,000 genes. Plants and animals, including humans, have more genes—about 25,000 for humans. Plant and animal DNA molecules are arranged as two paired strands in the familiar double helix form. A human cell has 46 molecules of DNA, each of which has 50 to 250 million bases.

Why can't two parents with blue eyes have a child with brown eyes?

Many aspects of your appearance are the result of heredity—traits that you inherited from your parents. Frequently, features like hair color, eye color, skin tone, height, and general build run in families. Siblings often have similar appearances. Although there can be great variations within a family, there are some combinations of traits that cannot occur. For example, two parents with blue eyes never have a child with brown eyes. Why is this so?

In the 1800s, Gregor Mendel determined the basic principles of inherited traits by studying how specific traits are passed from one generation of plants to the next. In organisms that reproduce sexually, offspring inherit many traits from the parents. Each organism inherits two possible characteristics for a given trait—one from each parent—and passes on one of those two traits to each of its own offspring.

If the two inherited traits are different, one of them is dominant, which means the gene that controls it is activated. The other gene for the trait is recessive. Although it is not activated, it can be passed on to the next generation.

The gene for brown eyes is dominant over the gene for blue eyes. That means that a person with a gene for brown eyes and a gene for blue eyes will have brown eyes. If both parents have blue eyes, then neither of them has a gene for brown eyes, so none of their children can have brown eyes. On the other hand, a person with brown eyes can have a recessive gene for blue eyes. If both parents have a dominant gene for brown eyes and a recessive gene for blue eyes, their child can inherit two genes for blue eyes. Therefore, two parents with brown eyes can have a child with blue eyes but two parents with blue eyes cannot have a child with brown eyes.

Ff Fast Facts

Inheritance is actually not quite as simple as one gene being dominant and the other recessive. Sometimes both genes can be activated at the same time. You can inherit a gene for straight hair or a gene for curly hair. A person with one of each gene is likely to have wavy hair, somewhere between straight and curly. To complicate things even further, some traits are controlled by a combination of two or more genes, each of which exists in the DNA in two forms, one from each parent.

Do all dogs belong to the same species?

A lion and a tiger are similar in size and if you look at their features, both are clearly cats. The two animals, however, come from two different species. A Great Dane and a Chihuahua are both dogs but one is many times larger than the other and their features differ from one another much more than the features of the two big cats. The two dogs, however, are considered to be from the same species. Why are very different dogs from the same species while similar cats represent different species?

Part of the answer lies in the definition of species. The original concept of species was introduced as an organizing tool when scientists thought that life on Earth fell cleanly into well-defined categories that were completely separate from one another. Unfortunately (or maybe fortunately), the natural world is not always so clear. Scientists do not completely agree on a current definition of species.

Ss **Science Says**

"I look at the term species as one arbitrarily given, for the sake of convenience, to a set of individuals closely resembling each other … it does not essentially differ from the term variety, which is given to less distinct and more fluctuating forms. The term variety, again, in comparison with mere individual differences, is also applied arbitrarily, for convenience's sake."

—Charles Darwin (1809–1882)

In the most familiar definition, a species is a group of organisms capable of interbreeding and producing fertile offspring. Other definitions look at the isolation of one group from another in nature, considering groups that would have no opportunity to mate as different species, even if they could physically produce offspring. Recently, more precise definitions probe similarities in DNA to classify relationships.

Practically speaking, a Great Dane and a Chihuahua cannot mate due to the extreme size difference. However, the Great Dane can mate with a smaller dog, such as a German Shepherd, which can produce offspring with a collie. The collie in turn could fall for a beagle, producing cute and fertile offspring. And who wouldn't want a beagle-Chihuahua mixed-breed pup?

There is always variation in a species: each zebra has its own pattern of stripes; polar bears vary in size within a certain range; each gorilla in a band has its own personality and ways of interacting with others. No other species, however, shows the wide range of sizes, shapes, and behaviors as you find with dogs. All of the dogs you know can trace their ancestry back to wolves, but today very few of them could ever be mistaken for a wolf. Why would this particular species be so varied compared to others?

In part, it's because dogs live in a wide environmental range. Wild dogs adapted to cold regions would have longer fur and bigger bodies than those from warm places. The major cause of the diversity, however, appears to be us—humans. People have been breeding dogs for at least 12,000 years. Many dogs were bred for specific reasons: chasing fleet game animals, following small pests into holes, detecting and warning against intruders. Within the last few centuries, appearance became a primary driver among dog breeders. Each breed has an ideal appearance that drives the pairings.

Ff   Fast Facts

Appearance in not the best indicator of how closely related one species is to another. To determine relationships between species, scientists must compare DNA—the more similar the DNA, the more closely the species are linked genetically. Although they all look and act in similar ways, DNA analysis has shown that all dogs are descendants of wolves, but not of coyotes or jackals. The coyotes and jackals are distant cousins.

Most of the dogs that you see around you owe their appearance, not to natural selection, but to an artificial selection process driven by humans. The species adapts by filling many different niches tied to human wants and needs rather than by adapting to pressures of the natural environment.

What was the purpose of the Human Genome Project?

The Human Genome Project (HGP), sponsored by the U.S. Department of Energy and National Institutes of Health, lasted for 13 years, studying the basis of human heredity—the entire set of genes that determines the characteristics of an individual. The goal was a complete map, showing the location of each gene on the human DNA molecules. What is the value of this map and how could it be used?

The HGP had several goals: to identify all of the genes in human DNA, determine the sequence of the approximately 3 billion bases, develop tools to analyze the vast amount of information, transfer the technologies to the private sector, and address

De **Definition**

An organism's **genome** is its complete set of DNA. The human genome contains about 6 billion base units. Most of the genome is identical for all humans, but a small part of it differs. Within these differences are the genes that make each person unique.

ethical, legal, and social issues. The mapping of the *genome* was completed in 2003 but the analysis will continue for many years.

Because each individual has different DNA, the map was made using samples from several individuals. It does not show the exact sequence from any one person, but identifies the general sequence of human DNA and the specific areas that correspond to genes.

Scientists expect to develop many benefits from knowledge of the human genome. Many diseases are either caused or affected by our genes. Each person has a unique sequence and the differences in those sequences can be correlated to the risk of developing certain diseases. One practical benefit of knowing which genes are related to diseases is the ability to determine which people are more likely than others to contract the diseases. Tests have already been developed to predict a genetic tendency to develop breast cancer, cystic fibrosis, and Alzheimer's diseases in some cases. Knowing that you have a genetic tendency toward one of these diseases can help determine treatments or prevention measures.

Us **Uncommon Sense**

Understanding the human genome is not a magic bullet that will allow us to understand the causes and prevent all diseases. Many diseases can be traced to environmental, not hereditary sources. Some diseases that appear to run in families may be caused by shared habits and environmental conditions, rather than genetic causes. Even in cases where specific genes are associated with a particular disease, for example, breast cancer, the genetic code generally only shows a tendency toward the disease. Usually other factors determine whether a person is affected or not.

Different people respond to drugs in different ways. Some drugs are extremely effective in certain people but ineffective or even dangerous to others. Knowing how genetic differences influence the way drugs interact with people's bodies could allow doctors to find the right drug and the right dose for treatment of a specific person.

Beyond determining the best drug for treatment, there is a possibility of applying genetic information to create and adapt drugs. Ultimately, genetic therapy may provide a way to cure, or at least manage, genetic diseases by making changes in a patient's DNA itself.

How does cloning work?

In 1997, Scottish scientists announced the results of an experiment in which they had developed a sheep named Dolly by manipulating cells in a laboratory. Unlike sheep with two parents, Dolly had genes that were identical to those of her mother. How can a new organism be created by cloning?

When we talk about cloning an organism, we mean creating a genetic duplicate of the organism—that is, one that has the same DNA. Although Dolly made a lot of news, cloning was around long before the appearance of the first cloned sheep. In nature, single-celled organisms reproduce by cloning when the cell splits into two new cells with identical genes. Gardeners love to make copies of a favorite plant by rooting a stem cutting. Identical twins start out as a single fertilized cell, which splits to form two genetically identical embryos—clones of one another.

Ff **Fast Facts**

Although articles about cloned animals often refer to them as identical copies, they are not exactly the same. The DNA that controls inherited characteristics is located in the nucleus, but there is also a second source of DNA, the mitochondria. The mitochondria are small structures in the cell that are involved in energy production. Mitochondria are always inherited from the mother because they are part of egg cells but not sperm cells. Mitochondrial DNA represents only a small part of the DNA in a cell. A clone created by nuclear transfer is actually not as closely related to its nuclear parent as identical twins are related to one another.

If you know any identical twins, you are probably aware that they are not identical. Twins can have very different personalities and, although their physical features are generally quite similar, they are not absolutely identical. It is their DNA that is the same, but many other factors influence individual people or individuals of any other organism.

Dolly, the sheep, however, was created by a different process with the same result.

Scientists removed the nucleus from an egg cell of a sheep, keeping the rest of the cell intact. They then isolated a single cell from another adult sheep and removed its nucleus. When they transferred this nucleus to the egg cell, they created a new complete cell with genetic information identical to that of the adult sheep. This cell then divided to form an embryo, which was implanted into the uterus of a surrogate mother. Dolly then developed into a clone of her parent, the donor of the nucleus.

Can humans be cloned to produce copies of people?

The advances in cloning technology that led to the creation of Dolly raised many questions about cloning that had not previously been discussed. One particular question—whether the same techniques can be used for humans—has raised a number of other technical and ethical questions. Is it possible to clone humans, and if so, should we do so?

Reasons that have been proposed for cloning humans include helping infertile couples produce genetically related offspring and bringing deceased relatives back to life. The process of cloning humans should be the same as that of cloning sheep, so biologists believe that humans could be cloned for these purposes, or others.

There are some ethical concerns, though, that do not necessarily apply to the cloning of animals (although not everyone agrees on application to animal cloning). Most of the time, cloning does not work. Many embryos die after being implanted in the uterus. In addition, a large percentage of the offspring die before, or shortly after, birth. Furthermore, many of the animals that survive suffer from defects in their hearts, lungs, or other organs and malfunctioning immune systems.

There is a second type of human cloning that avoids some, but not all, of the ethical issues. In therapeutic cloning, a nucleus taken from a person is inserted into a donor egg to form an embryo with that person's DNA. The embryo is not implanted into a woman's uterus to grow into a baby, but instead is allowed to divide several times to produce stem cells. The stem cells are removed and used to grow any type of tissue, which can be implanted to the donor of the original nucleus. Because they are genetically identical, these cells will not be rejected by the immune system, a major problem in organ or tissue transplants. Therapeutic cloning has been proposed as a possible cure for many diseases, including Alzheimer's and Parkinson's.

The main ethical concern of therapeutic cloning is the creation of an embryo specifically for destruction as the stem cells are harvested.

How does human DNA compare to animal and plant DNA?

All plants and animals have DNA that forms double strands. The organisms that are described by the DNA molecules are very different, though. How much difference is there in the DNA of different organisms?

In many ways, the DNA molecules of all organisms are the same. Bacteria, plants, spiders, and humans all have DNA that is coded with the same four bases. These bases provide a code for manufacturing proteins. Some of the processes that occur inside our bodies also occur in other organisms—even plants. For example, when we eat foods containing sugars from plants, the sugar molecules provide energy to our bodies. The same process breaks down the sugars inside the plant itself. Some of the genes used by plants to produce proteins used during the conversion of sugar to energy may be identical to the human genes that produce the same proteins.

Comparisons of the genomes of humans and those of several apes indicate that the chimpanzee is humanity's closest relative. A comparison of the genes in chimpanzee DNA and human DNA show that about 98.7 percent of the genes are the same. All of the other apes appear to be more distant relatives of the chimps. A study of the genome of chickens shows that we even share about 60 percent of our genes with them.

Ff **Fast Facts**

Is a mushroom more closely related to a marigold or a sparrow? This may seem like a strange question, but the answer is even stranger. While fungi and plants may seem to be closely related to one another, genetic studies have shown that fungi actually have a closer relationship to animals. It is true that the relationship is very distant, but certain DNA sequences and protein production in cells have shown that fungi are more like animals (including humans) than plants. This may explain why certain fungal infections are very tenacious. It is hard to find substances that will attack these fungi but not their hosts.

Why are some genetic diseases more common in men than in women?

During the eighteenth and nineteenth centuries, a number of European nobles suffered from a disease called hemophilia, in which blood does not clot normally. The disease can be very dangerous, particularly if injuries lead to uncontrolled internal bleeding. Hemophilia is a hereditary disease, caused by a defective gene. Hemophilia is common among all races. Children with hemophilia are almost always boys, although it is inherited from their mothers' families. Why would a disease affect one gender substantially more often than the other?

Hemophilia is an example of an X-linked hereditary disease. It is caused by a defective gene on a particular *chromosome* that differs by gender. Humans have 23 pairs of chromosomes, each pair containing one strand of DNA from each parent.

De | **Definition**

A **chromosome** is a single strand of DNA, packaged together with certain proteins. Every human cell, except for sex cells, has 46 chromosomes arranged in 23 pairs. Each chromosome has hundreds or thousands of genes. The sperm and egg cells in humans each contain 23 chromosomes, 23 pairs when the egg is fertilized.

However, in one of the chromosome pairs—the one that determines the gender of the offspring—the two strands do not contain the same number of genes. In the twenty-third chromosome pair, females have two completely matched DNA molecules. In males, however, the DNA strand from the mother is longer than the one from the father. The female chromosome is labeled as X and the male chromosome is labeled Y (based on their shapes as observed through a microscope).

In all other chromosome pairs, there are two genes and the dominant gene determines the person's characteristic. In males, all of the genes in one end of the chromosome pair are donated by the egg cell from the mother. If one or more of the genes on this segment are defective, the trait will be observed, even if it is recessive. This is because there is no corresponding gene from the father.

Several hereditary diseases are linked to the X chromosome, including hemophilia and color blindness. In general, these diseases appear in males. A girl will only have the disease if her father has it and her mother carries the recessive gene for it. Half of the boys whose mothers carry the gene will have the disease, regardless of whether their fathers have it or not.

How do DNA tests work?

Every crime show has a lab that performs a DNA test in order to find absolute evidence of the identity of the criminal. A sample is placed in the machine and it spits out the name of someone whose DNA is an exact match within seconds. While the results actually take much longer to obtain than shown on TV, DNA testing is a powerful identification tool. It has been used both to convict and to exonerate people accused of a crime, establish paternity with very close to 100 percent accuracy, and identify victims of accidents who could not be identified by any other means. How does DNA testing work?

Most of your DNA (in fact, more than 99 percent of it) is exactly the same as that of your parents, your neighbors, and even a random stranger from the other side of the world. There are some sections, however, that vary from person to person. It is this small fraction, spread throughout the total genome, that makes you the unique person that you are. No one else has the same sequence of bases in all of these sections of DNA unless you have an identical twin.

DNA analysis looks at some of these specific sections, called markers. A small sample of DNA is taken from cells found in body fluids, skin, hair follicles, or the inside of the cheek. After the DNA is isolated from the cells, millions of copies are made using an enzyme that speeds DNA reproduction. Other reactions then break the DNA molecules apart at specific locations to isolate particular markers. These markers are compared to the unknown sample.

Any one specific marker is shared by many people, but the chance of two people having two identical markers is much smaller. In an ideal situation, we would look at the whole DNA sample, comparing all the possible markers to make identification. With today's technology, though, this is not possible. DNA analysts look at a number of different markers to create a "DNA fingerprint." The more markers identified, the better the chances that the match between two samples is accurate.

Us | **Uncommon Sense**

A DNA test using a blood sample is no more accurate than a test using cells from a cheek swab. The swab does not collect saliva for the tests. It actually picks up loose cells that were shed by the tissue of the inside of the mouth. Because all cells contain the same DNA, the swab will provide the same DNA as a blood sample. If a hair sample is used, it must be pulled, not cut, because hair is not made of living cells. When a hair is pulled, however, it carries with it some cells from the living follicle.

Without looking at the whole of a person's DNA, it is not possible to absolutely identify a particular person as the source of a DNA sample. In the case of identical twins, it is never possible. However, by using many markers, the odds of a correct identification are extremely favorable. Normally, between 6 and 13 markers are compared. At the higher end of this range, it is extremely unlikely that an incorrect identification will occur, but not absolutely impossible.

On the other hand, it is easy to identify a mismatch in DNA analysis. If any of the markers is different, then the DNA does not match. In a criminal investigation, therefore, a suspect can be eliminated as the source of a sample based on the absence of a single marker. All DNA from a particular person matches from one end of each molecule to the other.

Because samples can remain useful for a very long time, DNA analysis has been used to reopen cases that were once considered solved. The results have far greater reliability than other forms of identification, including eyewitness testimony. A number of people have been cleared of crimes of which they had been convicted, sometimes decades earlier, by analysis of evidence held in storage.

Can DNA analysis be used for anything other than criminal investigation?

The most familiar (thanks to television) use for DNA testing is forensic analysis of crime scene samples for evidence to use in the identification of a criminal. While this is an important application of the ability to analyze DNA, it is only one of many investigations where DNA information is useful. In what other ways do scientists use DNA?

Even within forensic analysis, there are other ways to use DNA. In some cases, it is not only the suspect who must be identified but the victim as well. Identification of human remains if they are severely decomposed is a bit tricky. Many of the features we use day to day to tell one person from another, such as facial shape, are made of soft tissue and are destroyed over time after death. Dental records are often used but they are useful only when the jaws and teeth are available and when a record exists for comparison.

DNA analysis provides another tool for identification. Only a tiny sample of DNA is necessary for the analysis. Comparison samples can come from the possible victim's toothbrush, hairbrush, or other personal sources. Even when such a sample is not available, markers can be compared to those of close relatives. The largest single

DNA identification project took place after the attack on the World Trade Center in 2001. More than 1,600 victims of that disaster have been identified by DNA analysis of bodies and parts of bodies recovered from the scene.

Archaeologists rely on DNA to establish relationships among ancient cultures. Samples from archaeological sites have shed new light on human migrations, traced interactions among cultures, and identified mummies and other human remains. Information about the changes in DNA over time provides new insight into the origins of modern humans and the relationships between modern humans and earlier people, such as the Neanderthal people.

Of course, because DNA is a substance common to all life, DNA analysis is not limited to humans. There are an almost countless number of applications where it is useful to know relationships between living organisms or the source of materials in which DNA analyses are powerful tools. Examples include detecting bacteria and other organisms that cause pollution, tracing the source and development of diseases, and identifying poaching of endangered wildlife species.

Ff **Fast Facts**

Although it is not possible to build dinosaurs from ancient DNA, as shown in Jurassic Park, scientists have found that some DNA lasts a long time. Ice core samples in Siberia have yielded DNA from large mammals that lived 30,000 years ago and from plants that grew as long as 400,000 years ago. Although a few older samples have been reported, there is a possibility that the samples became contaminated over time and the DNA that was recovered was actually produced in more recent times.

How do paternity tests work?

Before the development of DNA testing, paternity testing often depended on matching of blood types. Because so many people share each of the blood types, however, the test can really only be used to eliminate the possibility of paternity rather than to identify the father. DNA paternity testing is much more accurate. How can DNA be used to determine whether people are related?

A paternity test is based on the fact that your entire DNA comes from your parents. One half of your DNA exactly matches that of your mother and the other half exactly matches that of your father. DNA samples are prepared from the child, the mother, and the possible father following the same procedures used to match samples in criminal investigations.

For a paternity test, however, we are not looking for an exact match. On average, half of the markers will be found in the mother's DNA. These markers are ignored. The other markers are compared to the DNA sample from the man who may be the child's father. If any of the markers do not match, then it is not possible that he is the father. The confidence of a positive result, in which all markers match, depends on the number of markers tested. In general, paternity tests identify enough markers to report a 99.9 percent probability that the positive identification is accurate.

"The time with which we have to deal is of the order of two billion years. What we regard as impossible on the basis of human experience is meaningless here. Given so much time, the 'impossible' becomes possible, the possible probable, and the probable virtually certain. One has only to wait: time itself performs the miracles."

—George Wald (1906–1997)

Chapter 12

Biology—Medicine and Health

"Public health is purchasable. Within a few natural and important limitations any community can determine its own health."

—Hermann M. Biggs (1859–1923)

The science of medicine has changed substantially in the past several centuries. The discovery of microorganisms as the cause of infection was one of the greatest advances. Knowing how diseases are caused and spread has led to improvements in hygiene and the development of vaccines to prevent diseases. Antibiotics allow patients to quickly recover from bacterial infections that would have been deadly a century ago.

Why do we need vitamins?

More than 2,300 years ago, scurvy had been identified as a disease that strikes sailors on long journeys. The body of a person suffering from scurvy has trouble producing collagen, a component of connective tissue. Its symptoms include bleeding gums, loss of teeth, severe joint pain, slow healing of wounds, and even death. British sailors became known as Limeys after it was discovered that eating citrus fruits on long voyages prevented scurvy. Today we know that people contract scurvy when they do not ingest enough vitamin C, one of a group of substances needed in small amount for the body to function properly. Why do we need vitamins?

A vitamin is an organic compound (a compound based on carbon) that an organism needs but cannot produce in sufficient amounts, if at all. Your body can produce one, vitamin D, in your skin—but only in response to sunlight. So if you live in an area with limited sunlight, you will need additional vitamin D. In the United States, vitamin D is generally added to dairy products. About one third of the vitamin K that you need is produced inside your body, although not by your own cells. Bacteria in the intestines produce the compound and you absorb it through the intestinal walls. Most of the vitamins that you need are available in a well-balanced diet, although sometimes supplements are helpful.

Vitamins play an important role in a wide range of functions within cells, tissues, and organs. Each vitamin has many different roles within your body, so a vitamin deficiency can have many symptoms, which do not always seem to be related. Vitamin A, for example, is important for vision because it is the only source for retinal, a compound required for the rods and cones in the eye to function, so lack of vitamin A is often associated with vision problems. It also plays a role in skin health and in the immune system, so a deficiency of vitamin A can also cause acne and a tendency to develop infections easily.

De **Definition**

An **enzyme** is a protein used by a living cell to control the rate of a chemical reaction. Many of the reactions inside the cell could not occur at body temperature without the help of an enzyme.

There are six vitamins in the group of B vitamins. Among other roles in the body, they work with *enzymes* to convert food into energy and to build new cells. Most of the B vitamins are found in fresh fruits and vegetables. Some of them are also available in meats and dairy products. Most breakfast cereals are fortified with B vitamins. Because they are involved in so many different functions, a deficiency of these vitamins can cause a variety of symptoms, including diarrhea, skin problems, headache, weakness, and nervous system disorders.

Most people know that citrus fruits contain a lot of vitamin C, but other good sources include broccoli, green peppers, and most fruits. Vitamin C is necessary for humans to produce collagen, a protein fiber that connects organs and tissues throughout the body and gives structure to cells. Signs of a deficiency of the vitamin include muscle and joint pain, loose teeth, and easily damaged blood vessels, which causes easy bruising and slow healing. Vitamin C also removes chemical substances that damage cells and provides protection against disease.

Vitamin D is necessary for the processes that control the distribution of elements, such as calcium and phosphorus, in your bones. One of the obvious symptoms of diseases, such as rickets, which are caused by a shortage of vitamin D, is weak or deformed bones.

Vitamin E protects your body from some of the aging processes by reacting with chemicals in your body that damage cells and increasing the flow of oxygen to cells. Your heart and lungs depend on vitamin E, as well as the red blood cells that carry oxygen throughout your body. When you do not get enough of the vitamin, these tissues break down and do not work as efficiently. Foods that provide vitamin E to your diet include nuts, whole grains, seeds, and spinach.

While some of the vitamin K that you need is made in your intestines, most of it must come from your diet from leafy green vegetables, meat, and dairy products. Vitamin K is necessary for strong bones and in the clotting of blood. Signs that you have too little vitamin K include a tendency to bleed or bruise easily, and brittleness in bone tissue.

Us   **Uncommon Sense**

Vitamins are good for you, so it may seem that taking more of them is better. It is possible, though, to get too much of a good thing. Vitamin C and the B vitamins dissolve in water, so your body easily gets rid of any excess. The other vitamins, however, tend to dissolve in fatty tissues in your body. If you ingest more of them than your body needs, they accumulate and can eventually become a toxin. Overdoses of the fat-soluble vitamins can cause insomnia, fatigue, and high blood pressure, among other symptoms. It is unlikely that you would overdose as a result of a normal diet, but you need to be careful not to take excess vitamin supplements.

Is cholesterol really bad for you?

We have all read warnings about too much cholesterol in our diets. Depending on what article you read, eggs, meat, and dairy products are either essential parts of your diet, hazardous to your health, or both. And then, you also hear about "good" cholesterol and "bad" cholesterol. So is cholesterol in your diet really bad for you?

Cholesterol is a waxy substance that your body uses when it builds cell walls. It is also necessary for the proper manufacture of vitamin D and some hormones. Cholesterol is manufactured in your liver and then distributed to cells in the bloodstream. The problem with cholesterol is that it does not dissolve in blood. Instead, it is carried by proteins.

There are two forms of protein-cholesterol complexes: low-density lipoprotein (LDL) and high-density lipoprotein (HDL). This is where good and bad come in. Although both forms are always present and are necessary for your body to function, an excess of the LDL form ("bad") can cause problems.

LDL carries cholesterol from the liver to the rest of the body. When there is too much of it in the blood, deposits form on the walls of arteries, making them less flexible and blocking the blood flow. If the arteries become too constricted, these blockages can lead to a heart attack or a stroke. On the other hand, HDL carries cholesterol from the blood back to the liver, where it is discarded. By carrying away and reducing the amount of cholesterol, HDL reduces the deposits on the arteries, making HDL the good guy. Within the normal range of total cholesterol, a high ratio of HDL to LDL reduces risks of heart attack or stroke.

Recent research indicates that limiting the cholesterol in your diet is not the most effective way to control the cholesterol in your blood, which is where it does the harm. In general, reducing dietary cholesterol has a relatively small effect on the blood concentration. For most people, about 75 percent of the cholesterol in the blood is made in the liver. If you have high cholesterol, even a small reduction is helpful, but there is a more effective way to attack the problem.

It appears that the best way to control cholesterol is to eat a diet that is low in specific types of fat. Saturated fats, which include most animal fats—meat, dairy, egg yolks—and some vegetable fats—coconut and palm oils—tend to raise cholesterol. Because they raise both the LDL and HDL levels, the overall effect of saturated fats is negative, so their consumption should be limited.

Us **Uncommon Sense**

Although high cholesterol levels are often associated with being overweight, people with any body type can have high cholesterol, particularly if they have a diet high in unsaturated and trans fats. Because there are no noticeable symptoms of high levels, you should have your cholesterol checked regularly.

Another form of fat, called trans fat, actually decreases HDL while increasing LDL in the blood. Trans fats are produced by the addition of hydrogen to the fat molecules in vegetable oils. These fats, found in stick margarine, many snack and processed foods, and commercial deep-fried foods (such as french fries) should be eliminated from the diet if possible because of their strong negative effects on cholesterol.

A third type of fat, unsaturated fat, actually reduces LDL levels while increasing HDL levels. These fats are found in most vegetable oils, including corn, sunflower, soybean, canola, and olive oils.

Do spicy foods cause stomach ulcers?

Stomach ulcers are potentially deadly holes in the stomach lining that can cause severe pain and internal bleeding. In severe cases, the hole can pass completely through the stomach lining and allow its corrosive contents to escape into surrounding tissue. Thirty years ago, if you had asked any doctor what causes stomach ulcers, the answer would include one or more of these factors: stress, spicy food, and alcohol. Today, you would get a completely different answer. If spicy foods don't cause ulcers, what does?

For the past century, the treatment for ulcers was bed rest, a bland diet, and antacids. These treatments helped reduce the discomfort but the results were generally not permanent. The pain returned when the treatment stopped. Recent research showed that stress, spicy foods, and alcohol can make the symptoms of ulcers worse, but they do not cause them. Most stomach ulcers are the result of a bacterial infection in the stomach lining. So what is the best treatment? Antibiotics cure ulcers by killing the bacteria with a month or two of treatment.

The bacterium that causes ulcers, *Heliobacter pylori*, was first discovered in 1982 by researchers in Australia, who later won the Nobel Prize in Medicine for the discovery. Up to that time, no one had expected any bacteria to be able to survive the acid conditions of the human stomach.

The hypothesis that ulcers was caused by an infection was not immediately accepted. One of the researchers, Barry Marshall, actually caused an infection and ulcer in his

Ff | **Fast Facts**

The discovery that ulcers are caused by microbes was a surprise to most medical professionals. Now research is underway to find out whether other common diseases, including arthritis and atherosclerosis (hardening of the arteries), may also be caused by infections.

own body, and then cured it with antibiotics to provide additional evidence. This is not a research practice that is generally encouraged.

After scientists knew to look for *H. pylori*, they found that about half of the people in developed countries, and almost everyone in some parts of the world, have the bacteria in their stomachs. The infection apparently begins in childhood and continues for life. However, only 10 to 15 percent of people with the bacteria develop ulcers. Researchers are now working to find out why some people develop these symptoms of infection while others do not.

How do bacteria become resistant to antibiotics?

When antibiotics were first introduced in the 1950s, they were a wonder drug. For the first time in history, many deadly diseases were under control. Infections that previously had been life-threatening could be brought under control in a few days and completely cured within weeks.

Unfortunately, some of the wonder has worn off. Bacteria have shown an ability to resist the effects of antibiotics. Some infections are barely controllable and some antibiotics that were once very effective are now practically useless. What causes antibiotics to become less effective?

For many bacteria, the inside of your body is the perfect habitat—warm and moist with a constant flow of nutrients. Many of them co-exist peacefully with you and some are even necessary for your body to function. Others are invaders who produce chemicals that are toxic or reproduce rapidly, taking nutrients needed by your cells. When these invaders are detected, the immune system reacts to destroy them.

Generally, the fight between bacteria and your immune system goes on, day and night, without your awareness. Sometimes, however, the bacteria get the upper hand and start reproducing out of control. Then your defense systems kick into high gear in an attempt to wipe out the invaders. Many of the symptoms of an infection—including swelling, high temperature, and redness—occur as the immune system fights the bacteria. If the bacteria reproduce too quickly, the infection can cause severe damage to tissues and organs, and even death.

Antibiotics help the immune system by killing the bacteria. They are poisonous to particular bacteria but not to the cells of your body. They interfere with the bacterial cell by disrupting normal functions, such as building new cell walls, blocking DNA synthesis, or disrupting its metabolism so that it cannot produce necessary compounds, such as proteins. As a result, the bacterium dies or is not able to reproduce and the infection and its symptoms go away.

Unfortunately, bacteria are sometimes able to avoid the effects of the antibiotic. There are several ways they can survive. Sometimes they change the structure of the membranes that allow chemicals to pass into and out of the cell, keeping the antibiotic compound from entering the cell. Another defense technique is changing the structure of an enzyme or other compound that is the antibiotic's target. Another way a bacterium can protect itself is to destroy the antibiotic before it can do any damage.

How do bacteria learn to protect themselves? Actually, it is not a learning process. What happens is that an individual organism develops immunity to the antibiotic. When it divides, its descendants are also immune. As the other bacteria around them are destroyed, this colony has more and more access to resources, so it thrives.

The initial resistance can come from a *mutation*, or change in the DNA of a single organism. If the mutation protects the bacterium from the antibiotic, it reproduces and an antibiotic resistant strain now exists.

Another way to get resistance is to pick it up from other bacteria. Antibiotics do not kill all types of bacteria, so there are always harmless species around that are immune. Sometimes a harmful bacterium can join with a bacterium of another species, mixing their DNA. In other cases, it can scavenge DNA remnants from dead bacteria. DNA that makes the antibiotic harmless can then lead to an antibiotic-resistant strain.

De Definition

A **mutation** is a change in the genetic sequence in the DNA of an organism. Mutations can be caused by errors during making DNA molecules or by exposure to radiation or chemicals. If mutations increase the chance of an organism's survival, they are likely to be passed on to future generations.

Us Uncommon Sense

Antibiotics are not the cure-all that they were once believed to be. They are only effective against bacteria, while many diseases are caused by viruses. One major reason for the development of antibiotic-resistant bacteria is the tendency, in the past, for doctors to prescribe antibiotics when they were not necessary. Overuse of antibiotics has led to many strains of bacteria developing resistance.

Why do flu epidemics seem to start in Asia?

Flu epidemics can travel rapidly through the population. We go to clinics to get shots to protect us from flu and to slow its spread. It seems as if there is a new epidemic predicted every year and new strains of the virus appear regularly. If you watch the news, it seems that these epidemics usually start in Asia. Do flu epidemics really start there, and if so, why does this happen?

Influenza, or flu, is caused by a virus that works its way into cells and takes over. Then, instead of operating normally, the cells begin to produce new viruses, which attack more cells. Many of the uncomfortable symptoms of flu occur as your body fires up the immune system to fight the invasion. Other symptoms—such as coughing, sneezing, and lung congestion—occur as dead cells, killed by the virus invasion, accumulate in tissues.

Generally, after your body has fought a virus once, the immune system "remembers" it and delivers a knockout blow quickly when the next invasion starts. Sometimes though, a small change in the virus structure can hide it from the system. Then your body has to learn to fight it all over again.

Ff   Fast Facts

Pigs are also a source of flu viruses that can infect humans. Both human flu viruses and bird flu viruses can live in their bodies, providing an opportunity to exchange genetic material. This leads to occasional outbreaks of "swine flu" in humans.

One of the reasons that flu can become such a problem is that there are many different strains of the virus. Some of them infect only people. Others infect animals such as birds, horses, and pigs. Normally these viruses are not a problem for people. However, viruses occasionally mutate and gain the ability to pass from one species to another. Influenza viruses have eight segments of genetic material, two of which determine whether the virus is able to infect a particular type of host. Viruses are able to swap genes with one another, so if human and animal viruses come together, a new virus can form.

Avian flu viruses occur naturally in bird populations around the world. Although wild birds carry the viruses, they usually do not get sick from them. The viruses are very contagious, though, and domestic birds, such as chickens, ducks, and turkeys can contract flu from the wild birds.

Mutations in viruses and transfer of DNA among them can occur at any place in the world. So why do so many of them seem to come from Asia? For viruses related to birds and other farm animals, part of the reason may be related to differences in farming around the world. In Europe and North America, poultry are generally raised in large farms. In much of the rural parts of Asia, though, most households have some chickens, ducks, or other birds that provide eggs and meat. Therefore, a much larger part of the population has close contact with domestic birds. Another reason that Asia is often a source of flu viruses may be even more basic. A very large percentage of the world's population lives in Asia. That means there are a lot more people who could be the original host to a mutated virus.

How do vaccines prevent disease?

Smallpox is a devastating disease. During the eighteenth century, it killed about 400,000 people each year in Europe alone, and left many of its survivors disfigured or blind. Worldwide, hundreds of millions of people have died from smallpox infections. In 1796, Edward Jenner discovered that people who had contracted a related but much milder disease, cowpox, tended to be immune to smallpox. He used this observation to develop a vaccine to protect against smallpox infection. In 1979, smallpox vaccination finally reached enough of the world's population that the disease was declared to be extinct. How do vaccines work?

Ss **Science Says**

"Diseases can rarely be eliminated through early diagnosis or good treatment, but prevention can eliminate disease."
—Denis Burkitt (1911–1993)

When you catch a disease, from a virus or bacterium, your immune system produces *antibodies* to fight the disease organism. After you recover, your body remembers how to make the antibodies for that particular disease. If you are exposed to the organism again, the system kicks in, produces the right antibodies, and wipes out the invader before it can become a problem. That's why many diseases, such as chickenpox, generally only occur once during your lifetime. In the past, these diseases were known as childhood diseases because most people contracted them early in life and were then immune to the disease.

> **De** **Definition**
>
> An **antibody** is a protein used by the immune system to identify and disable a specific bacterium or virus. Antibodies are produced by white blood cells. The general structure of all antibodies is similar, but there are differences in small regions of the molecule that allow millions of antibodies to exist, each of which is matched to a specific target.

Vaccines are a way to teach the immune system to produce the necessary antibodies without ever having had to fight an actual infection. There are several ways to do this. The smallpox vaccine exposed people to organisms that could cause a similar disease, whose symptoms were much milder. Even if the person contracted the disease, it did not create the problems of a smallpox infection. The organisms were similar enough, though, that the antibodies designed to attack the vaccine also worked against the smallpox virus. Some vaccines are produced by weakening the disease organism so that it cannot reproduce rapidly and cause a severe infection. The immune system responds to the foreign organism and designs antibodies to fight it.

Vaccines can also be produced by killing bacteria or deactivating viruses with chemicals or radiation. These organisms are now harmless and they can be injected into the body without any danger of causing a disease. Their presence can, however, cause the body to fight them as if they were living organisms, and design antibodies to destroy them.

Can sugar pills really stop pain?

When researchers test a new drug to determine its effectiveness, they need to compare the results with those for people who do not take the drug. Usually, though, they have to also compare to people who think they are taking the drug but are not actually doing so. Doctors have known for a long time that it is possible to treat some conditions just by convincing the patients that they are receiving an effective treatment. One way to do this is to give a pill that looks like the real drug but contains no medication. Is it really possible for these "sugar pills" to stop pain and cure diseases?

Research has shown that, in some circumstances, a fake treatment, called a placebo, can be as effective as real medicines. For example, 30 to 40 percent of patients with conditions ranging from high blood pressure to arthritis and even Parkinson's show improvement after taking placebo pills. Fake surgery, in which small incisions were made on patients' knees, exhibited the same results as actual arthroscopic surgery in one study of arthritis treatments. How can a placebo have this kind of effect?

No one really knows exactly what happens when a placebo has a positive medical effect, but it appears to tap into a built-in healing power of the brain. For example, your brain can make chemicals that are similar in their effects to morphine. Studies have shown that, under certain conditions, these chemicals are released by the brain if the patient believes that a pill will relieve pain. The pain relief is real, not imagined, because the brain chemicals have the same effect as pain killers that are known to work.

Ff **Fast Facts**

Researchers compare effectiveness of a proposed drug with a placebo in order to determine how effective the medicine really is. If it is not more effective than the placebo, it has no real effect, even if it is better than no treatment at all.

In one study of the effectiveness of aspirin in preventing heart attacks, the aspirin was so much better than the placebo that the study was stopped long before the five-year test was completed. The effectiveness was so clear that doctors began to recommend regular doses of aspirin for all at-risk patients.

Placebos do not always work, though. Unlike those conditions where the brain apparently has untapped resources, diseases such as cancer do not respond to placebos. Even so, there is an amazing range of problems which can apparently be treated by giving your brain permission to just go do its thing.

Why do joints get sore and red when you have arthritis?

Arthritis can be very painful and cause joints to swell, feel hot, and turn red. These symptoms, known as inflammation, are similar to the symptoms of some infections, even though arthritis is not caused by bacteria or viruses. What causes inflammation in joints?

In many cases, the symptoms of arthritis are similar to those of infection, and in fact have the same cause. Inflammation is the process that your body uses to protect itself from an invasion by bacteria and viruses. The immune system sends in white blood cells and other tools to destroy the invaders. In addition to the white cells, a number of proteins and other chemicals are sent to the site to provide protection.

In some diseases, such as bursitis and some forms of arthritis, the immune system goes into defense mode when there are no foreign cells around. When this happens, the immune system attacks the body's own tissues, treating them as if they were an invasion themselves. This type of inflammation is known as an autoimmune response.

Uncommon Sense

Arthritis is not a disease of the elderly. Although many older people do suffer from arthritis and from joints that have been damaged by the disease, people of any age can have arthritis. Juvenile arthritis is a joint inflammation that strikes children under the age of 16.

The symptoms of inflammation caused by an autoimmune response include redness and swelling as extra blood is pumped into the region. This swelling, along with action by some of the immune system chemicals, causes stiffness in the joint and stimulates nerves, creating painful sensations. If the inflammation is not reduced, the increase of cells and other substances in the joint can damage the joint itself by causing swelling in the joint lining. The swelling can also damage cartilage that provides padding between bones.

Does the color of a bruise indicate how bad it is?

When you bang you arm sharply against a cabinet or bang your shin against a chair, you often get a bruise. It may appear to be a bit of reddening at first, but over time bruises can take on many different colors. Does the color of a bruise indicate anything about its severity?

Fast Facts

Sometimes a bruise does not go away on its own. Instead of breaking down the trapped blood, the body walls it off, forming a firm, sometimes painful, swelling between the skin and the muscle. This swelling, called a hematoma, may need to be drained by a doctor before it goes away.

A bruise occurs when small blood vessels under the skin rupture or tear. The blood that flows out of them becomes trapped in a pool under the skin and can't return to normal circulation. In general, bruises are not a major health problem, even though they are unsightly, and they tend to go away on their own. Bruises are usually caused by a bump or fall. Some people, especially older adults, bruise very easily, so they may not even notice or remember what caused the bruise.

The colors of a bruise do not indicate how bad it is, but instead are more closely related to its age. In the first few minutes after you get a bruise, it will generally be red or pink as blood collects beneath the skin. Within a few hours, the layer of blood takes on the typical black and blue shades.

Over the next two to four weeks, the bruise slowly heals. As it does, the body breaks the trapped blood down into its components, which are then recycled. Various compounds in the blood have different colors, as do the compounds into which it breaks down. As the bruise heals, various components and breakdown products show through the skin. In general, the dark black, blue, or purple of a new bruise gradually fades to be replaced by a succession of color. Violet gives way to green or dark yellow, which gradually fades to light yellow and then disappears.

Why is it so hard to find a cure for the common cold?

A few centuries ago, people were pretty much at the mercy of nature when it came to disease. No one knew that many diseases were caused by microorganisms. There were no antibiotics to cure an infection and no vaccinations to prevent one. Many of the terrible infectious diseases of the past can be controlled fairly well today. Most of the worst can be prevented by immunization and hygiene. When an infection does occur, it can often be treated with antibiotics. One curse remains untouched, though—the common cold. Why can't we find a way to prevent or cure colds?

Many virus infections are prevented by vaccination but, unfortunately, that doesn't work for the cold. As it turns out, there is not just one common cold—there are hundreds of common colds. A whole series of viruses, known as rhinoviruses, cause the symptoms that we know as a cold. You probably have a natural immunity to some of them, thanks to your immune system's reaction to previous colds, but there are more to come. Beyond that, due to mutations of the existing viruses, there are probably new cold viruses coming out all the time.

If you can't stop the virus, how about treating the cold to cure it faster? There have been quite a few proposed cures for the cold, including nose sprays, zinc, and vitamin C. Unfortunately, none of them has held up too well in research to determine their effectiveness.

Actually, most of the symptoms of a cold are not caused by the virus anyway. Sneezing, a runny nose, fever, and congestion are caused by your immune system's efforts to fight the virus. Inflammation in your sinuses does not really do too much to get rid of the cold viruses, but your body includes it.

For now, your best bet for preventing a cold is frequent hand washing to prevent transferring viruses to your mouth and nose. The best treatment is plenty of fluids and rest. The traditional chicken soup is probably as good as anything else you can come up with. It doesn't fight the virus either, but it tastes better than most of the alternatives.

"When are we going to say cancer is cured? I'm not sure when that will happen, if that will happen because cancer is a very slippery disease and it involves a vast number of cells in the body and those cells are continually mutating."

David Baltimore (1938–)

Part Q

Earth and Space Sciences

Now we get to the bigger picture. Earthquakes, hurricanes, and volcanic eruptions have both fascinated and terrified people throughout history. If anything is evidence of capricious gods, these should be the prime examples. Yet even Earth follows rules of nature—nothing moves without a force that pushes; nothing is capricious after all. Geologists and meteorologists try to answer the how and why of the world around us, while environmental scientists look at the systems of living and nonliving parts that support life on Earth.

For the even bigger picture, astronomers look out from our home into space. The biggest questions of all are those posed by cosmologists—the size, age, history, and fate of the entire universe.

Chapter 13

Geology—The Ground Below

"Scientists still do not appear to understand sufficiently that all earth sciences must contribute evidence toward unveiling the state of our planet in earlier times, and that the truth of the matter can only be reached by combing all this evidence. ... It is only by combing the information furnished by all the earth sciences that we can hope to determine 'truth' here, that is to say, to find the picture that sets out all the known facts in the best arrangement and that therefore has the highest degree of probability."

—Alfred Wegener (1880–1930)

Standing on top of a mountain, you can feel the solid earth beneath your feet and as you look out over the valley, you can imagine it as a stable, unchanging view. But that is an illusion. The earth is a dynamic, constantly changing planet. The solid rock beneath your feet was once the bottom of the sea or a molten mass far below the surface of the planet. Where you look out over an unchanging valley, there may have once been a mountain twice as tall as the one on which you are standing.

As strange as it may seem, that solid earth is constantly moving. The continent on which you are standing is rushing toward a collision with another continent; or maybe the collision is already underway. Continents may move only a few inches each year, but on the scale of the planet's 4 billion years, things are rushing along. Geology is the science of these changes.

Why does Earth have a magnetic field?

From the earliest times, mariners traveling beyond the sight of land have needed a reliable direction indicator to keep them on course. The stars work very well, but only if you can see them. For more than a thousand years, sailors have used the compass to keep them on course, either along or in combination with astronomical observations. The magnetic field of the planet provides that constant directional reading, available day or night, clear or cloudy. Even today, with GPS systems to provide instantaneous, reliable readings of position and direction, every ship carries a compass as a backup. Why does Earth have a magnetic field in the first place?

Magnetism is a force generated by a moving electric charge. As an electric current flows through a wire, it generates a magnetic field that can be detected by bringing a magnet close to the wire.

The largest magnet on Earth is the planet itself. Although it is impossible to sample material from deep inside the Earth, geologists have evidence that its core is made of a mixture of iron, nickel, and small amounts of other metals. The inner core is a solid metal ball surrounded by a liquid metal layer, which rotates slightly faster than the layers above it. The rotation churns the liquid, causing flowing currents of liquid metal. Under the conditions in the moving liquid, some of the electrons become separated from their atoms. As a result, the moving electric charges in the liquid generate a magnetic field.

Because it is generated by currents in a liquid, Earth's magnetic field is not constant. The North and South Poles are not located exactly at the poles of Earth's axis of rotation. In fact, the magnetic North Pole is currently about 600 miles from the true North Pole and is moving at about 25 miles per year. If it continues to move at the same rate and direction, the magnetic pole will travel from its current location, north of Canada, into Russia in the next half century. Navigators using compasses must make corrections for the distance between true north and magnetic north.

Ff   Fast Facts

The planet Venus is about the same size as Earth and is believed to have a core whose composition is similar to Earth's. However, Venus does not have a magnetic field because it rotates on its axis once every 243 Earth days instead of once per day. The observation that Venus does not have a magnetic field is one piece of the evidence that currents in the flowing liquid part of the core is responsible for Earth's field.

The strength of the magnetic field is also variable. It has weakened by about 10 percent since it was first measured in the middle of the nineteenth century. Geological evidence shows, however, that this is still twice as strong as the average value over the last million years.

Us   Uncommon Sense

Earth's magnetic field reverses direction, causing the North and South Poles to switch, at irregular intervals averaging about 380,000 years. Several movies have presented such a switch as devastating to life on Earth. In reality, the direction of the poles has reversed hundreds of times without any evidence that such reversals have caused mass extinction of living species.

Why is Earth's interior hotter than its surface?

The deepest mines in the world extend more than 2 miles below Earth's surface. One of the challenges to miners digging gold and diamonds in these deep tunnels is heat. Two miles underground, the temperature of the rock walls measures greater than 130°F, requiring air-conditioning and protective clothing. Much deeper than these mines, the temperature is so high that rocks melt into a molasses-like consistency. This molten rock reaches the surface as fiery hot lava. Why does the temperature of Earth increase with depth?

The main source of energy on the surface is sunlight. Stronger sunlight means a higher temperature. When you enter a cave in the side of a mountain, you feel cool air because light cannot penetrate the mountain and heat the air inside. So it would seem as you go deeper underground, farther from the sun's warmth, temperatures would drop.

If the sun were the only source of heat, that intuitive reasoning would work. However, there are other sources of heat inside Earth, so as you travel deeper, the temperature rises. At two miles, it is uncomfortably warm. At three miles, mining would almost certainly have to be done by remote control robots as temperatures approach 160°F. At the planet's core, the temperature is estimated to reach as high as 8,000°F (5,000°C), nearly as hot as the surface of the sun.

Ff Fast Facts

The depth beneath the crust to reach very hot temperatures varies from place to place. There are a few areas (generally the same places where you find volcanoes) where the Earth's crust is relatively thin. Geothermal energy becomes very economical in those places. In the state of California, there are 33 geothermal power plants that tap into Earth's interior heat. A district heating system in Reykjavik, Iceland, uses hot water from beneath the surface to heat 95 percent of the city's buildings.

There are two sources of heat inside Earth. About one-third of the heat is left from the formation of the planet. The current theory is that everything around us condensed from a giant cloud of gas more than 4 billion years ago. As gravity pulled matter into a ball, its pressure increased and its volume decreased. Along with those changes, there was a huge increase in temperature as the matter was compressed. The energy that caused this temperature increase gradually leaks into space from the surface, but the layers of material between the core and the surface act as an insulator, so the loss of energy occurs fairly slowly.

Ff Fast Facts

Unlike Earth, the moon does not have any active volcanoes. It is much smaller than Earth, so the moon's surface area is much larger compared to its volume and it radiated most of its heat into space long ago. Heat from radioactive decay is also lost too rapidly to keep the mantle molten. Jupiter's moon, Io, however, has more volcanoes than any other body in the solar system, even though it is about the same size as our moon. Io orbits very close to its giant parent and its interior is heated by the motion of tidal forces.

The second, and larger, source of energy that heats Earth's insides is the breakdown of radioactive elements. This is the same energy source used for nuclear power plants. The nucleus of a radioactive atom breaks apart to form two or more smaller nuclei. When this happens, a tiny amount of the atom's mass is converted to a lot of energy. Some radioactive elements slowly break down over billions of years, producing the energy that constantly heats the interior of the planet. Therefore, even though the interior is insulated from solar energy, it is much hotter than the surface.

Why do Africa and South America look like they fit together?

Looking at a map of the world, you will likely see (as have countless geography students in elementary schools) that the continents look a bit like the pieces of a jigsaw puzzle. In the sixteenth and seventeenth centuries, explorers sailed across the face of the planet drawing maps as they traveled. Mapmakers back home noticed that South America and Africa seem to have complementary coasts as if they had been cut apart. Is this apparent match just a coincidence, or were these two continents once connected?

This question bugged Alfred Wegener, a German meteorologist in 1912. The fit between these continents, as well as other seeming matches, seemed too good to be a coincidence. In addition, evidence of ancient glaciers in Africa and tropical climates in North America indicated that the continents must have moved.

Wegener also found further evidence that the fit between continents is not an illusion. Some fossils of ancient land animals on the West Coast of Africa and the East Coast of South America indicate that the two continents, although now thousands of miles apart with very different ecosystems, once hosted identical populations.

Wegener proposed a hypothesis—continental drift—that all of the continents were once joined into a supercontinent, which he called Pangaea. Unfortunately, Wegener was unable to suggest a mechanism that would explain how something as large as a continent could move from one place to another. It was not until the mid-1960s that geologists were able to do so. According to the modern theory of plate tectonics (see Chapter 2), which is built on Wegener's ideas, it is no coincidence that South America and Africa look as if they could fit together. They are actually two parts of a broken continent.

Ss Science Says

"The Wegener hypothesis has been so stimulating and has such fundamental implications in geology as to merit respectful and sympathetic interest from every geologist. Some striking arguments in his favor have been advanced, and it would be foolhardy indeed to reject any concept that offers a possible key to the solution of profound problems in the Earth's history."

—Chester R. Longwell (1887–1995)

Why do so many earthquakes occur along the Pacific Coast?

On October 17, 1989, people around the country sat in front of their televisions, ready to watch the third game of the World Series matchup between the San Francisco Giants and the Oakland A's. Just before the game was to start, a major earthquake shook San Francisco—the first nationally televised earthquake. The Loma Prieta earthquake was the worst to hit San Francisco since the devastating earthquake of 1906. In between, however, the city had experienced hundreds of smaller quakes. Why are there so many earthquakes along the western coast of the United States?

De Definition

The **epicenter** of an earthquake is the point on the surface that is directly above the earthquake focus, the place where the original motion of the rocks occurs.

Earthquakes occur when rocks beneath the surface suddenly shift, releasing stresses that have built up over time. A tremendous amount of energy is released in seconds, moving the ground above. A really strong quake can be felt hundreds of miles away from the *epicenter* of the earthquake.

To understand why the West Coast has so many earthquakes, you have to know how the stresses on the rock build in the first place.

As the plates of Earth's crust move, they constantly bump into one another. Because of the mass of rock involved, these collisions have a lot of energy. For example, the collision between the subcontinent of India and Asia has built the Himalayan Mountains. Although the motion of the plates is slow (2 to 12 centimeters per year), the mass is so great that massive earthquakes occur where plates collide.

If two plates move past one another, a lot of the energy builds as stress when huge masses of rock scrape against one another. This stress builds until the rocks suddenly slip. This slippage causes a sudden release of stress that may have been building for hundreds of years. The jolt shakes and shatters the ground nearby and, sometimes, very far away.

Ninety percent of earthquakes occur along the boundaries between two moving tectonic plates. As it turns out, the western coast of North America includes two plate boundaries. From Oregon to Alaska, the North American Plate is colliding with the Juan de Fuca Plate, making that region an active earthquake area. Along the California coast, the Pacific Plate is moving northwest relative to the North American Plate. The San Andreas fault stretches about a thousand miles along the boundary. As the plates grind against one another, the rocks along the fault slip and jolt along, creating one earthquake after another, including the Loma Prieta earthquake in 1989.

Ff **Fast Facts**

Although many earthquakes occur in California, where two plates are slipping past one another, the two strongest earthquakes to be recorded in the United States occurred in Alaska, where two plates are colliding, in 1964 and 2002. The strength of an earthquake depends to large extent on the length of the fault that shifts during the quake. During the Loma Prieta earthquake, a 25-mile-long fault shifted during a 7-second period. During the earthquake that destroyed much of Anchorage, Alaska, in 1964, a 600-mile fault shifted over a period of 420 seconds.

How do fossils form?

Much of the information that we have about the history of life on Earth comes from studying *fossils*. Every major museum of natural history has a collection of dinosaur bones; at a shale or limestone quarry, you can often find rocks that contain shells of sea creatures, even far above sea level; sort through a pile of coal and you will likely

 **De** **Definition**

A **fossil** is the preserved remains or traces (such as footprints) of plants, animals, or other organisms.

find the impression of ancient leaves (and get plenty dirty at the same time). How did these fossils form and show up where we find them?

Fossils are not all formed in the same way. Generally, when people think of a fossil, dinosaur bones come to mind. These fossils form when the animal dies and is buried before its body decomposes or is eaten by scavengers. Generally, only the hard parts of the body are fossilized. That's why we see displays of dinosaur bones. As the flesh decays, water and minerals penetrate the hard parts of the body—bones, shells, teeth, claws, and such.

The sediment that settles around these remains preserves them and minerals harden the organic material. The body part is buried by sediment which is eventually converted to rock under the pressure of layers above it. The fossil is buried until someone digs it up or it becomes exposed by erosion of the rock around it.

Often the original material decays after time and is replaced by minerals that form a hard rock in the shape of the original bone. This is how petrified fossils, such as the ancient trees in a petrified forest, are preserved.

Us **Uncommon Sense**

The fossil record does not include all of the organisms that once lived on Earth. Some organisms do not have hard parts, so they are much less likely to be preserved than others. Other organisms may have died in places where fossilization is unlikely so they decayed or were eaten by scavengers. Organisms living in water, for example, are much more likely to be preserved than organisms living on land. The records provided by fossils are only a small picture of life on Earth through time.

Another type of fossil that can be found in sedimentary rocks is a trace fossil. As sediment forms something is pressed into it, leaving an impression. Think about how small children make handprints in clay as a present for their parents. The clay hardens, leaving a permanent record of a tiny hand. In the same way, something that disturbs sediment can leave an impression that later dries and hardens. As other layers fill the impression, they do not destroy the trace and it becomes a permanent feature that remains as the rock forms. Trace fossils can include an impression of soft tissue, skin, and even the footprints of ancient animals. Fossilized footprints of prehistoric humans have also been discovered.

Not all fossils are found in rock. If you have ever seen a piece of polished amber with an insect embedded in it, you were looking at a fossil. The insect died thousands or millions of years ago when it was trapped in the sap of a tree. As the sap dried and hardened, the insect was preserved as a fossil. The La Brea tar pits in Los Angeles hold many animals that were trapped thousands of years ago in the tar and sank into the viscous black liquid. Because the tar prevented decay, the soft tissue was preserved.

How did the Colorado River make the Grand Canyon?

Standing on the edge of the Grand Canyon is one of the best ways to really see the power of natural forces and time. It forms a huge gash across the desert, almost 300 miles long, more than a mile deep along much of its length, and as much as 18 miles wide. Can the Colorado River really have formed this canyon? How long did it take?

Looking at the walls of the canyon, you can see layer after layer of rocks. These are sedimentary rocks, formed as sediment collected at the bottom of ancient oceans and seas. Near the bottom of the canyon, these rocks are almost 2 billion years old, while the rocks at the top were formed "only" about 200 million years ago. It took a long time, half the age of the planet, to forms these deep deposits.

It did not take nearly that long, though, to make the canyon. About 75 million years ago, the North American tectonic plate started to slide over top of another plate, raising the plains of the Colorado Plateau by 5,000 to 10,000 feet. This collision is also forming the Rocky Mountains to the east of the plateau.

About 5 million years ago, an opening was formed from the plateau to the Gulf of Mexico. The elevation change from the higher reaches of the plateau to the sea caused the water to flow rapidly, carrying away sand and rock. During the ice ages, the flow of water increased and the river cut into the rock very rapidly.

Ss | **Science Says**

"With the sole guidance of our practical knowledge of those physical agents which we see actually used in the continuous workings of nature, and of our knowledge of the respective effects induced by the same workings, we can with reasonable basis surmise what the forces were which acted even in the remotest times."

—Giovanni Arduino (1714–1795)

Even today, as the water runs downward during a period of heavy flow, it picks up loose rocks and boulders, some as large as a car, and carries them along. This debris helps cut into the sides of the canyon, making it wider, and into the river bed, making it deeper. Because the desert has few plants to stabilize the soil and rock, erosion is very rapid on the plateau. On the scale of geological change, the Grand Canyon was formed in the blink of an eye.

Where does lava come from?

A volcanic eruption is one of nature's most impressive shows. Red-hot lava flows out of the ground at temperatures as high as 3,600°F (2,000°C). Sometimes, as with Hawaii's Kilauea Volcano, it flows slowly into the sea, causing the water to boil and evaporate into huge clouds of steam. In other volcanoes, such as Mount St. Helens in Washington, it causes the mountain to blow its top, sending boulders flying for miles and raining hot rock from the sky. Where does this lava come from?

Lava is made of molten rock from beneath the Earth's crust. Inside the planet, it is called magma, and it is a mixture of liquids, gases, and solids. The Earth's crust, which includes all the land and the oceans, ranges from about 5 to 50 miles thick. Beneath the crust is the mantle, which is made mostly of very hot, solid rock that is somewhat fluid, like modeling clay. This mantle is about 2,000 miles thick, surrounding the iron-nickel core of the planet. At the top of the mantle, the pressure on the hot rock is low enough that it becomes liquefied in a layer that is about 60 miles thick. The tectonic plates of the crust float on this layer, known as the asthenosphere.

Although the pressure is lower than that on the layers beneath it, the magma of the asthenosphere is still pressurized compared to the surface above it. When there is an opening in the crust, magma can squirt out to the surface. This normally occurs at the boundaries between tectonic plates, but sometimes there is an opening in the middle of a plate. One such opening has allowed magma to flow into the center of the Pacific Ocean, building the Hawaiian Islands.

Ff ▶ Fast Facts

Most of Earth's active volcanoes occur along the boundaries between tectonic plates. More than half of the volcanoes that are above sea level are part of the Pacific Ring of Fire, a string of volcanic mountains that extends along the western coasts of the Americas, the Aleutian Island chain, and along the eastern coast of Asia, including all of Japan and the Philippines, Indonesia, and many of the islands of Oceania.

The difference in types of volcanic eruptions depends on the composition of the magma, in particular the amount of gas dissolved in the molten rock. Lava from magma that has few bubbles flows gently to the surface. Researchers (and sometimes tourists) safely approach the lava.

Magma that is full of gas bubbles, on the other hand, can create a truly spectacular show. As the magma rises toward the surface, the gas bubbles grow. If a layer of rock holds the pressurized magma in place, pressure can build to the point of eruption. The pressurized gas blows the top off the mountains, spewing hot lava as high as 2,000 feet into the air. These eruptions can cause great devastation, sometimes hundreds of miles away. The greatest eruptions can send so much material into the sky that it shades the surface from sunlight. This can disrupt weather patterns around the world for years.

How does crude oil form underground?

It's hard to imagine the modern world without petroleum products. Petroleum fuels our vehicles, paves our roads, and is the basic raw material for products ranging from drugs to plastic to building materials, and the search for it fuels political campaigns and wars. Petroleum was practically unknown before 1850, but it has been one of the primary sources of energy since the early part of the twentieth century. Now, depending on whose estimates you trust, we may or may not have enough to last another century. How did oil get into the ground in the first place?

Petroleum (literally, rock oil) does not actually *get* into the ground; instead, it is made there. It is formed by the decomposition of dead organisms, primarily marine plankton and algae. As they die, their bodies sink to the bottom of the sea and mix with mud and other sediments. Because of the dead organisms, the sludge at the bottom of the sea is rich in materials that contain carbon and hydrogen. Over millions of years, layer after layer of this material settles. New deposits put pressure on the older deposits and the temperature rises as a result of pressure and decomposition of the organic material.

Under the influence of this high temperature and pressure, the carbon-containing materials react to form long chains of carbon atoms attached to hydrogen atoms. A mixture of these *hydrocarbon* molecules in various lengths makes up the thick brown or green gunk that we call crude oil, or petroleum.

De Definition

A **hydrocarbon** is a compound whose molecules are made up of carbon and hydrogen atoms. Fossil fuels, such as petroleum, natural gas, and coal consist of different types of hydrocarbons and some impurities.

Once the petroleum forms, it tends to float upward because it is lighter than the saltwater that was also trapped in the sludge. It rises until a layer of dense rock traps it, forming the oil deposits that our drills seek. This is a continual process, of course, because plankton and algae are still dying and sinking to the bottom of the world's oceans. The catch, though is, that it takes millions of years to convert them into oil. Now might be a good time to reconsider how we are using the oil that remains.

Us Uncommon Sense

Reservoirs of oil are not like underground lakes. The oil deposits are spread throughout cracks and pores in rocks such as sandstone and shale. As oil is pumped out, it flows through the rock layer, but much of the oil is trapped in small pockets. Drillers inject brine into the rock so that oil will float upward toward the well opening.

Why do some layers of rock in a cliff run up and down instead of side to side?

As you drive along the road in a mountainous area, you can often see cliffs or road cuts where the layers of rock that form the mountains are clearly defined. Unlike the neatly stacked layers of rock seen in the Grand Canyon, the layers of rock in mountains are often jumbled. You can see bends and folds in the layers. Sometimes you can trace a layer for a great distance and then suddenly come to a break. The layer appears to continue, but the continuation is many feet above or below the original layer. Then there are places where the rock layers are stacked side by side in vertical rows. Why are these rock layers not neatly stacked, one on top of the other?

The layers of rocks that you see on the side of a cliff or road cut were once laid out in horizontal sheets, but something moved them—the same force that causes earthquakes. As tectonic plates collide, the rocks that they are made of are pushed and compressed. Imagine what happens to the metal of an automobile in a head-on collision. It bends and folds, taking shapes that look like the rock of the road cut.

Unlike a car crash, the collisions of continents take millions of years. Two kinds of bending occur. If the collision is slow and steady, generating heat in the rock and steady pressure, grains of minerals in the rock slide past one another and the layer of rock folds. Then there are no obvious breaks. The layers bend up or down, forming curves. Occasionally the rock is bent far enough to form hairpin shapes.

If the collision occurs fast enough to create a sudden shock (keeping in mind that sudden, in geological terms, may mean over many thousands of years), the rock layers can't bend as they do under a slow, gentle pressure. Then the rock layers break, forming a fault instead of a fold. In a fault, rock layers may be sharply tilted. You can often see faults as a break in layers, looking like the rocks slipped downward after breaking.

"We're looking at Earth science, observing our planet. Also space science, looking at the ozone in the atmosphere around our Earth. Also looking at life science. And on a human level, using ourselves as test subjects."

—Laurel Clark (1961–2003)

Chapter 19

Meteorology and Hydrology– Wind and Water

"What is clear is that the atmosphere is a continuous mass resting on the earth and the sea, and that these two react upon each other. Any disturbance which appears at any one point must make itself felt at very considerable distances from that point. We shall often have to seek for the cause of a certain phenomenon in another which has taken place perhaps in another hemisphere ... we have found interesting simultaneous relations between the barometrical pressure and the rain at different centres of action. So we have shown that there exists a sort of compensation between certain neighbouring centres of action."

—*Hugo Hildebrandsson (1838–1925)*

Although most of Earth is made of rocks and metals, these materials do not have as much effect on our daily lives as the air and the water around us. The weather determines what you wear each day, whether you stay inside or go out and sit in the sun at lunchtime, how much you will pay for heat this winter. Not only that, weather gives us something to talk about during those awkward silences.

Temperature and precipitation, the two main components of weather, are part of the movement of energy around the planet. Energy from the sun is absorbed and converted to heat. This heat sets wind and water into motion, redistributing energy from warm places to cool places—and making the weather. A worldwide system of wind and water currents has been described as a conveyor belt, transferring energy from one place to another.

Why does the jet stream flow from west to east?

Next time you fly between the coasts, check the schedule of your flights. It is very likely that the east-west flight is at least an hour longer than the west-east flight. The reason is that the flights get into the jet stream. Going westward, the plane is fighting a strong headwind; going eastward, it is pushed along by the same wind—the jet stream. What is the jet stream and why does it blow from west to east?

There are actually four jet streams in Earth's upper atmosphere, two in the Northern Hemisphere and two in the Southern Hemisphere. The jet streams are like broad rivers of air that move at a higher velocity than the air around them. These air flows are caused by a combination of differences in solar heating at different latitudes and the rotation of the planet.

The strongest jet streams occur at roughly 50 to 60° north and south of the equator (in the Northern Hemisphere this is around the U.S.–Canada border) and a weaker jet stream at roughly 30° from the equator (about the U.S.–Mexico border).

Because of Earth's tilt, the amount of heating of the atmosphere, ocean, and land decreases with distance from the equator. The air that is warmed in lower latitudes tends to rise in the atmosphere (here's that connection between temperature and density again). It then moves northward as cooler air moves from areas closer to the poles. The jet streams occur where warm and cool air masses come together.

This is where the *Coriolis effect* comes in. Because the planet is rotating on its axis, any point on the surface has a velocity from east to west that depends on its latitude. At the equator a point moves about 25,000 miles each day. On the other hand, at 60° north or south, a point travels only about half that distance in a day. The air above the equator moves with the surface at about 25,000 miles per day. As warm air from the equator moves northward, its momentum tends to keep it moving at the same speed eastward. However, the land beneath it is not moving as fast, so the air travels to the east faster than the surface. The winds are deflected eastward. The Coriolis effect explains the direction of the jet streams.

Jet streams form where the warm and cool air masses meet in the upper atmosphere. Their average speed is about 90 mph, but they can exceed 300 mph. The winds are strongest in the winter when the temperature difference between the equator and polar regions is the greatest. Although their channels can extend for thousands of miles in an east-west direction, the jet streams are generally only a few hundred miles wide (north-south) and they extend from about 4 miles to about 8 miles above the surface. They do not run in a constant channel but can extend far north or south of their average location.

Jet streams are an important part of the wind system that distributes solar energy on Earth. They tend to push weather systems around, having a great effect on local weather. Jet streams are also believed to play an important role in the paths followed by hurricanes and other tropical cyclones.

De  **Definition**

The **Coriolis effect** is an apparent deflection of an object moving in a straight line due to a rotating frame of reference. It is generally used to describe the tendency of moving water in the ocean and air in the atmosphere to appear to turn right when moving northward and to move left when moving southward.

Us **Uncommon Sense**

Despite the observations of Lisa Simpson, the Coriolis effect does not determine the direction of the whirlpool in a sink. While it does play a role in tropical cyclones and large ocean currents, the effect is much too small to be noticeable on the scale of the drain. The direction of the drain whirlpool is determined instead by the shape of the basin and by currents in the water, and it is equally likely to be clockwise or counterclockwise in either the Northern or Southern Hemisphere.

Why is it usually cooler at the top of a mountain than at the bottom?

When you think of the tops of tall mountains, you tend to think cold. Even mountains near the equator and mountains that rise from bases in hot deserts can have a year-round cap of ice and snow. Why are mountain tops so cold?

Remember that heat is transferred from one material to another by collisions between atoms and molecules. In air, nitrogen and oxygen molecules are in constant motion, colliding with one another and with surfaces with which they make contact. If the molecules are very energetic and the collisions are very frequent, the air feels warm. On the other hand, if the molecules have less energy and there are fewer collisions, the air is cooler.

Ff   Fast Facts

Death Valley in California, with an elevation of about 300 feet below sea level, is the hottest place in the Western Hemisphere. The average summer high temperature is 98°F. Meanwhile, at the 14,497-foot summit of nearby Mount Whitney, hikers have to cross snow and ice, even in August, to reach the summit.

At the base of a mountain, the air is heated as the ground absorbs solar energy as light and radiates it into the air as infrared radiation. Air molecules absorb this radiation, gain energy, and move faster. The temperature rises. As the air becomes warmer, it becomes less dense and rises.

However, the pressure of the atmosphere decreases as the elevation increases. That is because there is less air above it pushing down on it and compressing it. As air rises in the atmosphere, its pressure decreases, and according to the gas laws, so does its temperature. As you climb a mountain, the temperature, on average, drops about 5 to 6°F for each thousand feet of elevation change. Of course, that means that temperature increases as elevation decreases. If you descend into the Grand Canyon, you will find that it is 20 to 30°F hotter at the bottom than at the top.

Why do some clouds look white and others look gray?

Clouds are made of drops of water or crystals of ice that form when water vapor cools and condenses in air that rises from the surface. Clouds do not always have the same appearance when you look at the sky. Some clouds are thin, white wisps; others form a gray layer that covers the sky from one horizon to the other; in between, there are the big, fluffy puffs that are often white on top and gray on the bottom. Why do some clouds appear white while others appear gray?

As water evaporates from the surface of rivers, lakes, oceans, and even the soil, water molecules mix with the nitrogen and oxygen of the air. When a mass of air is heated by the sun or has contact with the warm ground, it begins to rise above the denser, cooler air around and above it. The air mass rises, expanding with the dropping pressure, and the entire mass cools. As the water molecules lose energy, they begin to condense and form tiny droplets or crystals.

The water droplets and ice crystals in a cloud reflect and scatter the sunlight that strikes them. The scattered light looks white, which is why the fluffy clouds that spread across the sky appear white. In fact, all clouds look white in the daytime when you look down on them from an airplane due to the refection and scattering of the light.

The amount of light that passes through a cloud depends on the density of the water droplets in it. High, wispy cirrus clouds contain very little water, so most sunlight passes through them. In some clouds, however, such as the nimbus clouds often associated with thunderstorms, the density of water droplets is much higher. In addition, from top to bottom, these clouds are usually much thicker. As a result, they reflect much more sunlight upward, away from the observer on the ground. The thicker the cloud and the more condensed water it holds, the darker the cloud appears. That is why the darkest clouds are usually associated with rain.

Ff **Fast Facts**

Cloud seeding is the process of spraying tiny particles of silver iodide in the upper part of a cloud to induce rain. The idea is that ice crystals grow around a particle, such as dust (or silver iodide) and then fall, melting to form rain. Unfortunately, even if it does rain, it is hard to be certain that it would not have done so without the seed.

Why don't hurricanes form near the North and South Poles?

The 2005 hurricane season was one of the worst on record. Altogether, there were 28 tropical storms that year. Four of them topped the scale of hurricane strength at category 5, including the devastating hurricanes Katrina and Rita. With just a few exceptions, these storms started in a band of the Atlantic Ocean between the latitudes 10°N and 30°N, roughly the northernmost extent of South America and the boundary between Florida and Georgia. Why don't hurricanes form farther north?

> **De** **Definition**
>
> A **tropical cyclone** is an intense circular storm that originates over warm tropical oceans. Cyclones demonstrate low atmospheric pressure, high winds, and heavy rain. Some tropical cyclones are called hurricanes or typhoons, depending on where they originate.

Hurricanes are *tropical cyclones*, which form over the warm waters of the Atlantic Ocean or the eastern part of the Pacific Ocean. Almost all hurricanes start out in a band of water within about 2,000 miles of the equator because that is where the specific conditions exist that lead to hurricane formation. Cyclones also form in the Pacific and Indian oceans. They are not called hurricanes, but they are the same type of storm.

The main ingredients that go into the birth of a hurricane are solar energy and the rotation of Earth. During the summer, the ocean waters of the tropical and subtropical oceans absorb a lot of energy from the sun. This energy heats the water near the surface of the ocean to temperatures of 80°F or more. Cyclones form only in areas where the top 150 feet or so of the water is heated to at least 80°F. As the molecules of water gain energy, more and more of them escape from the water's surface to become water vapor in the atmosphere above the water.

As this warm air rises, it begins to cool and condenses to form clouds. The condensation process releases a lot of energy, heating the air again and increasing its pressure. As the higher-pressure air moves outward, winds begin to build. Wind can move out in any direction. Because of the motion of the rotating Earth, air that is flowing north tends to be deflected eastward. The result is a counterclockwise circular motion of the air around the center of evaporation and condensation. The Coriolis effect causes a clockwise motion in the Southern Hemisphere.

As evaporation and condensation continue, a wind pattern develops, forming a circular flow around a cylinder of cooling air. As long as the wind pattern is over warm water and no other winds disrupt their flow, the cycle builds as more energy is added and the wind spirals faster and faster. When the wind speed reaches 74 mph, the storm is considered to be a cyclone. These storms reach at least 50,000 feet into the atmosphere and can be hundreds of miles across.

As long as the hurricane is above warm water, the cycle of evaporation and condensation can continue to feed energy into the moving air. Hurricanes rapidly lose strength when they move over cool water or land.

Much of the damage of a major hurricane that hits land is caused by the storm surge. Although a storm surge is sometimes portrayed in movies as a giant cresting wave, slamming into tall buildings and knocking them down, in reality, the storm surge is a slow, building phenomenon. Winds blow the water against the shore, where it piles up and overruns the land like a huge tide. The water can rise more than 20 feet during the storm surge, stranding ships as much as a mile inland.

What is the "wind chill factor" that the meteorologist discusses on the TV weather report?

As you watch the weather forecast on a cold, windy day, the announcer warns you to bundle up because the wind chill temperature is –5°F. However, when you look at the thermometer, you see that the temperature is actually +20°F. Why are the two temperatures so different?

The wind chill factor is a measure of the effect of the apparent temperature felt by exposed skin. The principle behind it is that moving air carries heat away more effectively than still air. Whenever the temperature of the air is less than your body temperature, heat radiates away from your exposed skin. However, if the air is not moving, the heated air tends to remain close and provides some insulation.

The wind chill factor is not a measure of actual temperature, so it does not affect the measured temperature of the air itself or of objects such as water, ice, or a thermometer. It only applies to radiating bodies such as people and animals.

A wind chill temperature (WCT) of –5°F means that your body experiences the same rate of heat loss as it would in still air at –5°F. If the WCT is –18°F or lower, exposed skin can experience frostbite in less than 30 minutes.

The concept of wind chill was first developed in the 1940s. In 2001, the National Weather Service issued an improved method for determining WCT based on advances in science, technology, and computer modeling that have improved the ability to measure the effects of temperature and wind. The complex formula used today is based on a model of the human face and skin interactions.

National Weather Service Wind Chill Values

Wind (mph)	Temperature					
	30	20	10	0	−10	−20
5	25	13	1	−11	−22	−34
10	21	9	−4	−16	−28	−41
15	19	6	−7	−19	−32	−45
20	17	4	−9	−22	−35	−48
25	16	3	−11	−24	−37	−58
30	15	1	−12	−26	−39	−53
35	14	0	−14	−27	−41	−55

frostbite in 30 minutes	**frostbite in 10 minutes**

Ff **Fast Facts**

The wind chill factor applies only in cold weather. A related concept used in hot weather is the heat index, which combines air temperature and relative humidity to determine an apparent temperature. High humidity decreases the rate of cooling by evaporation of sweat, so humid air feels hotter than dry air. The heat index is only used when the actual temperature is greater than 68°F.

Why is there often a strong wind just before a thunderstorm?

An oncoming thunderstorm often announces its approach. Rumbles of distant thunder may echo while the sky still looks clear. Then dark clouds build, towering high above you, with flashes of lightning. A very strong, cool wind often rushes just ahead of the storm, sometimes having enough force to knock down trees. How does this wind form?

A downburst is a strong wind that blows in all directions from a storm. In rare cases, downbursts have tornado-strength winds, up to 150 mph, and produce damage that is similar to tornados. However, the wind from a downburst differs from a tornado because it blows in one direction and it has a different cause.

A downburst is caused by the storm itself, as air flows downward from an area inside the storm.

In general, a downburst forms when there is an area of relative dry air within the storm. As rain falls into this air from above, it begins to evaporate rapidly. Evaporation is an endothermic process, which means that it absorbs energy, in this case from the air around the evaporating water. As the energy is absorbed, the air cools, making it denser. The density of the cold air causes it to descend toward the ground. If the air mass falls very quickly, it is diverted as it hits the ground, causing a strong wind that blows in every direction from the base of the storm.

Most downbursts extend less than 2½ miles, creating a strong wind that lasts for only a few minutes. Because the downburst extends in every direction, it can also be experienced as a wind blowing away from the storm after it has passed. They are particularly dangerous in aviation due to the downward component, known as wind shear, that can cause an airplane to lose altitude very rapidly.

Ss Science Says

"Part of the success of the invasion of the French Coast [in WWII] came by virtue of the fact that the weather forecast for that event, made by the American forces, was so inconceivably bad that the German meteorological experts, who were substantially better, simply couldn't believe that we would be so stupid as to make so bad a forecast, and could not believe that we would act upon it, and therefore could not believe that the invasion would occur at the time when it actually did. So, rather curiously, we profited by the bad state of our meteorology at that moment."

—Warren Weaver (1894–1978)

Why is the ocean salty while many lakes are not?

Take a mouthful of water while swimming in the ocean and you know right away that there is a difference between ocean water and the water you find in rivers and most lakes. Ocean water has a strong salty taste. If you don't rinse your swimsuit, it dries with a white crust of salt. Why is the ocean so salty when most other water is not?

Salts are compounds consisting of ions of a metal element and ions of a nonmetal element or a combination of elements. On average, about 3.5 percent of the weight of ocean water is salt.

Most of the salt in the ocean is sodium chloride, familiar as table salt, but altogether there are at least 72 elements found in seawater, most of them at very low concentrations.

The *salinity* of the ocean occurs because of a combination of water's ability to form solutions of salt compounds and the cycling of water into and out of the atmosphere.

De ❙ Definition

Salinity is a measure of the quantity of dissolved salts in water, generally expressed in parts per thousand (ppth: 1 part per thousand = 0.1 percent). Fresh water has a salinity that is less than 5 ppth; brackish water 5 to 29 ppth. On average, the salinity of the oceans is about 35 ppth. The salinity of the Dead Sea is about 315 ppth (31.5 percent salt).

When the oceans first formed, most likely during ancient volcanic eruptions releasing water from inside the planet, they were not nearly as salty as they are today. However, the atmosphere at that time, billions of years ago, contained a mixture of emissions from volcanoes, nasty things like hydrogen chloride (hydrochloric acid), sulfur dioxide, and hydrogen bromide. These chemicals all dissolve in water very rapidly.

Other materials were components of rock that dissolved in the acidic mixture after it fell from the sky. These materials include sodium, which combines with chloride ions to form sodium chloride. Other elements extracted from rocks and minerals include magnesium, calcium, and potassium.

Over time the ocean has continued to get more and more salty. Most of the salts that reach the oceans remain there while water constantly evaporates and recycles as rain. Rainwater dissolves more salts from minerals on land and carries them to the sea. Although the freshwater of rivers carries a low concentration of salt compared to the water of the sea, this salt then remains in the ocean, adding a bit more salinity.

Why are lakes less salty than oceans? Actually, while it's true that the salinity of most lakes is less than that of the oceans, there are exceptions. The concentration of salt in the Great Salt Lake in Utah, for example, varies between 5 percent to more than 20 percent compared to the average of 3.5 percent for the ocean. The salinity of Lake Superior is much less than 0.01 percent, typical for many freshwater lakes. This low value can be explained by the flow of water through the lake and into Lake Huron. On average, the water in Lake Superior is replaced over a period of 200 years, so the salt does not accumulate over time.

Ff **Fast Facts**

If the salt could be removed from the oceans and spread across the land surface of Earth, it would form a layer more than 500 feet thick, almost half the height of the Empire State Building.

Why does Seattle get twice as much precipitation as Spokane?

In an average year, the city of Seattle, Washington, receives about 38 inches of precipitation. The annual precipitation in Spokane, about 250 miles to the east of Seattle, however, is only about 17 inches. Why is the amount of precipitation so different in two cities that are so close together?

If you drive between the cities, or even look at a good map of the state, you will find a prominent feature between Seattle and Spokane—the Cascade Mountains. This mountain range runs in a north-south direction through the entire state. The mountains also provide a divider between areas of high precipitation and areas of low precipitation.

The prevailing winds across North America blow from west to east. The air reaching Seattle contains a lot of water vapor due to evaporation of water from the Pacific Ocean. As a result, the climate is relatively rainy. However, as this moist air reaches the Cascade Mountains and is forced upward, it expands due to lower pressure and its temperature drops. (If this seems to be a familiar theme, that's because it has such a major impact on the weather.) At the lower temperature, water condenses to form rain. The western slopes of these mountains tend to receive a lot of rain.

As the air passes beyond the mountains, it carries with it much less water vapor than it had before. In fact, as the cool, dry air drops closer to the ground, it warms, and water that is already in the ground tends to evaporate into the air, making the area even more arid.

Us Uncommon Sense

There is a good chance that in a science class, or even on a TV weather report, you have been told that it rains when air cools because cold air "holds less water" than warm air. Some textbooks have even shown the atmosphere as a sponge being squeezed to cause rain. This is not true because air does not "hold" water at all. It is not even correct to speak of air being saturated.

In reality, water in the atmosphere is constantly condensing and evaporating. When the temperature of the water vapor drops, the energy of the molecules decreases so there is more condensation than evaporation and water droplets form. The air itself does not play any part in the process besides carrying the molecules of water along with it.

Why can palm trees grow in Dublin but not in Portland?

The city of Dublin in Ireland is much farther north than Portland, Maine. Both cities are close to the Atlantic Ocean, yet Dublin's climate is warmer. In fact, the average low temperature in January in Dublin is 38°F and in Portland the average is 9°F. It is even possible to grow palm trees in Dublin. Don't bother trying that anywhere in Maine. Why are the climates of the two cities so different?

As strange as it may seem, the difference is attributed to the warm waters of the Gulf of Mexico. The Gulf Stream is an ocean surface current, sort of like a river in the ocean. A constant flow of water loops through the Gulf. A map of the ocean surface temperatures shows warm water, heated by the sun in the fairly shallow Gulf of Mexico, flowing out of the gulf and around the tip of Florida. Because this water is warmer than the water below it, it stays near the surface of the ocean.

The Gulf Stream current carries the water along the eastern coast of North America and it even passes Portland, Maine. However, the wind generally blows from west to east, so the warm waters off the coast do not affect the climate of Portland.

After leaving the coast of North America near Newfoundland, this still-warm water flows into the Atlantic Ocean and becomes the North Atlantic current, which flows near the western coasts of Ireland and Scotland. Here, however, it can affect the climate.

The winds blowing over the North Atlantic also tend to move from west to east, so they blow right across the surface of this warm current. The air is heated and moistened, carrying warmer, wetter weather to Dublin.

The Gulf Stream is part of a system of ocean currents that surrounds the globe. Warm water flows across the top of the ocean in surface currents. In the polar regions, however, water loses energy to the colder atmosphere. As the fluid seawater cools, it becomes denser and sinks toward the bottom of the sea. Cold-water currents deep in the ocean cycle water from the poles toward the equator. Where these currents come back to the surface, they bring cool, nutrient-rich waters from the deep ocean. One example of this phenomenon occurs off the coast of northern California. Unlike the waters of the Gulf of Mexico, which can reach 90°F in shallow areas, the Pacific Ocean waters off the shore of northern California remain at a chilly 50°F or so year-round.

Ff | **Fast Facts**

The Gulf Stream carries about 100 times as much water as all the rivers on Earth combined, flowing at speeds as high as 75 miles per day. The first person to describe and map the Gulf Stream was Benjamin Franklin, who was trying to explain why it took ships longer to travel from Europe to America than to travel from America to Europe.

Do you get wetter by walking or running in the rain?

Our final weather question is an old one. You are getting ready to go across the parking lot to your car, when a sudden spring shower pops up. You can walk to the car or you can run (you only have a small purchase to carry). Which tactic is best for reaching the car with minimal drenching?

In a thoroughly unscientific study of a parking lot during a rain, you will probably find that most people run for the car. Is that, however, the best option? If you walk you are in the rain longer, so it would seem that you get wetter. However, if you run, you collide with raindrops more often, so you may get wetter by running.

A British study in 1995 determined that the worst option (in terms of getting wet) is standing still in the rain, but that neither walking nor running offers a significant advantage. However, two meteorologists at the National Climactic Data Center in North Carolina—Thomas Peterson and Trevor Wallis—were a bit skeptical. For one thing, the study was a mathematical exercise, not a real experiment, and besides, their own calculations gave a different result.

The only way to be sure was to do an actual experiment. On a rainy day, the two scientists dressed in identical clothing and went out to answer the question definitively. One of them dashed 100 meters in heavy rain, while the other took his time, walking to the finish line. Weighing the clothes afterward, they found that the runner had absorbed 40 percent less water than the walker. So the question is now answered. As any child heading from the school bus seems to know instinctively, you don't get as wet if you make a fast dash for the door.

"Unfortunately no one can tell you with any scientific basis whether New Orleans will be hit this year, and if they are what strength the storm or hurricane would be. The best I can tell you is that New Orleans has always been an accident waiting to happen when it comes to hurricanes and flooding rains. This year is no different than in past years."

—*Steve Lyons (1955–)*

Chapter 15

Ecology and Environmental Science

"We have come to look at our planet as a resource for our species, which is funny when you think that the planet has been around for about five billion years, and Homo sapiens for perhaps one hundred thousand. We have acquired an arrogance about ourselves that I find frightening. We have come to feel that we are so far apart from the rest of nature that we have but to command."

—*Marston Bates (1906–1974)*

Ecologists and environmental scientists study different aspects of organisms on Earth. Ecology is the study of interrelations of living organisms and how they affect one another within a system of living things. Environmental science is a broader study of living systems, or environments. Environmental scientists use concepts of geology, chemistry, meteorology, agriculture, and many other fields, along with biology. In addition to studying the relationships between living things and their environments, environmental science addresses the human influence on living systems.

Why do forest managers use controlled burns?

Smokey the Bear began warning us about forest fires in the 1940s, and has been doing it ever since. His message has changed a bit over the years, however. At the time Smokey first appeared, the goal of forest management was to prevent all fires and to put out any fire that did start. Today, the National Park Service and other forest agencies start hundreds of fires every year. Why would the message about forest fires change over time?

De **Definition**

An **ecosystem** is a community of organisms—plants, animals, microorganisms, fungi—that interact with one another, along with the physical factors that make up their environment. Physical factors include soil, water, light, and temperature that are interrelated with the organisms.

Ecologists have learned that fire is an important part of many *ecosystems*. Since the first forests developed on land, fires have been started by lightning, volcanoes, and even by the decay of dead plant material. We now know that many species depend on fire in order to thrive, or even to survive and reproduce. In addition, many uncontrolled fires have been made worse, in terms of damage to human and natural resources, by earlier policies of preventing all fires.

The benefits of fire to an ecosystem include improvement to wildlife habitat, control of tree disease, removal of invasive species, and removal of fuels that can cause a fire to burn out of control if they are allowed to accumulate. In fact, the seeds of some species of conifers, including the sequoia, will not germinate unless they have been activated by the heat of a fire.

Even so, fires are not always desirable. Many forests and grasslands have changed substantially since they were allowed to burn freely. Human habitations and other structures are now an integral part of many areas where natural fires once burned. Controlled or prescribed burns are now used extensively to manage forests. In a controlled burn, excess fuel is often removed to prevent a fire from burning too hot. The fire is started when wind direction and speed make it easiest to control and crews stand by to prevent the fire from burning beyond the planned boundaries. Fire has become an important tool in maintaining forest health.

In the summer of 1998, the driest in the history of Yellowstone Park, fires burning out of control affected more than one-third of the park. Most of the fires were caused by lightning. Although human life and property were protected by a massive fire-fighting effort, there was no way to prevent the spread through the forest.

The Park Service was widely criticized for allowing natural fires to burn to the point that control was lost. Many news reports detailed the "destruction" of the park. Ecologists, however, saw a unique opportunity to study the effects of a large fire on a forest. Their research found that the fire did not destroy the forest and that it was, in fact, healthier within a few years than it had been before the fire.

Among their observations:

◆ Lodgepole pines are growing, reclaiming many areas where they had disappeared.

◆ Aspens have begun reproducing at faster rates than before as the forest floor was cleared.

◆ Grasslands recovered in several years with healthier plants due to increased nutrients in the soil.

◆ The amount of habitat available to some animals, such as bluebirds, has increased.

Smokey's message is not obsolete, however. Although controlled burns are used extensively, uncontrolled fires destroy many homes every year. They can also increase erosion when protective ground cover is removed prior to heavy rains. The message is still: "Only you can prevent wildfires!"

Ff Fast Facts

Many plants that have adapted to succeed in environments that include periodic fires cannot continue to exist without fire. Many wild grasses have very deep roots that easily survive a fire that destroys plants that compete with them. When there are no fires, the grasses of natural prairie ecosystems are replaced by bushes and trees and the prairie no longer exists.

Do flies serve any useful purpose?

On the whole, houseflies seem to be designed to be annoying. They buzz around the room, making noise at a grating frequency, and land on your body and on your food. It is possible that they can transmit a number of diseases and maggots, their larvae, are disgusting. Do flies serve any useful purpose?

Ecosystems are made of many organisms that depend on one another. It makes more sense to talk of the role of an organism in an ecosystem rather than the purpose of the organism. Although flies seem, on first glance, to have no value, they play important roles in nature.

A female housefly deposits about 100 to 150 eggs on something that can provide food for the larvae that will hatch from the eggs. This food is decaying material, such as garbage, animal droppings, or grass clippings. Between 8 hours and 2 days later, the maggots hatch and begin to feed. Eventually they form pupae and change into adult flies, restarting the cycle. Fly larvae are efficient disposers of garbage and other dead matter. Along with bacteria and other decomposers, they convert the material into other forms. Imagine how dead plant and animal materials and animal and human wastes would accumulate if there were no way to destroy them.

Throughout their life cycle, flies also play another role in their ecosystem—food for other organisms. Foraging insects, lizards, and small mammals feed on fly eggs, larvae, and pupae. Fish and other aquatic organisms feed on flies throughout their life cycle, as do many birds and other land animals. Although flies may be an annoying part of the world around us, they would be missed if they did not exist.

What is biodiversity and why is it important?

One of the goals of conservationists is to maintain the biodiversity of an ecosystem in order to protect the health of the system. What is biodiversity and why is it important?

Biodiversity refers to the variation in life forms within an ecosystem, or on Earth as a whole. There are three parts to biodiversity in a system:

Us   Uncommon Sense

You may often see the term ecosystem used to refer only to the "natural" world, excluding humans. Humans are actually part of the ecosystems in which they live. For example, your home is an ecosystem that includes people, houseplants, bacteria, dust mites, houseflies, a spider or two, and maybe a mouse. Each part of the ecosystem interacts with other parts.

- Genetic diversity is the variability in the genetic makeup of individuals in a particular species.

- Species diversity is the variety of species within an ecosystem.

- Ecological diversity is the variety of biological communities (e.g., forest, desert, grassland, tidal marsh, lake) that interact with one another.

Biodiversity at all levels is important for the general health of a system and the different species within it, including humans. Every species of living thing depends on other species for its existence. Herbivores need a variety of plants to provide all their needs; carnivores need a variety of prey species; plants depend on animals for pollination, seed distribution, and nutrients—the list of interactions is too extensive to list.

From a human viewpoint, biodiversity is important to provide for human needs. All of our food comes from plants or animals in our environment. At least 40,000 species are used by humans for food, shelter, clothing, and building materials. Diversity provides different materials for specific needs. Most of the medicines that we use come either directly or indirectly from a chemical compound that occurs in nature. We have only explored a small fraction of the available compounds that could be medically useful. In addition to material uses, the biodiversity of the planet provides us with leisure and cultural activities.

In a population of a single species, genetic diversity is a protection against disease. Individuals are able to react differently to different genetic diseases or infections. The diversity of species within an ecosystem protects the system in the same way. Some species are susceptible to specific stresses and attacks, while others are resistant. By having a wide range of species, an ecosystem can adapt and resist threats.

Ff **Fast Facts**

Extremophiles carry biodiversity to its limits. They are the organisms that live in conditions that would destroy almost all other life. Most extremophiles are microbes that have developed mechanisms to deal with severe environmental conditions. Examples of environments inhabited by extremophiles include geysers, in which water is near boiling; deep inside Antarctic ice; cracks in rocks miles below the surface; inside the human stomach; and the Dead Sea.

The value of biodiversity is most evident when a catastrophe strikes a monoculture (nondiverse environment). The Irish Potato Famine of the nineteenth century occurred because the population relied overwhelmingly on one crop for sustenance. When disease destroyed that crop, people were unable to adapt by turning to another source of nourishment.

Tropical rainforests are among the most diverse ecosystems. Tall trees reach high above the ground for access to light. Other species below them have adapted to lower light levels. Many of these plants and animals use the tall trees for shelter and food. At the ground level, still other plants live in low light conditions, obtaining nourishment from material dropped by plants and animals above them. On and under the ground, fungi, bacteria, and other decomposers break down leaves and other materials, making their nutrients available to the next generation of plants.

Ss Science Says

"Man will survive as a species for one reason: He can adapt to the destructive effects of our power-intoxicated technology and of our ungoverned population growth, to the dirt, pollution and noise of a New York or Tokyo. And that is the tragedy. It is not man the ecological crisis threatens to destroy but the quality of human life."

—René Dubos (1901–1982)

Why were phosphates removed from most detergents?

Soaps and detergents do not always break down in the environment and their residues cause problems in sewage treatment plants and waterways. In the 1950s many streams and rivers had an almost constant covering of foam that was toxic to many of the small organisms necessary for the health of the aquatic ecosystem. Materials were added to detergents to make them *biodegradable*. Other materials were added to make detergents work better in hard water. Several phosphorus-containing materials served both of these purposes. If they were useful in the 1950s, why have phosphates now been removed from detergents?

De Definition

Biodegradable means that a material can be broken down by biological organisms called decomposers. Most decomposers are bacteria or fungi.

While phosphorus-containing compounds in the detergents helps with their performance and biodegradability, phosphates in the water can also create problems. The breakdown of detergents by bacteria, which was one of the desired results, created an oversupply of an essential nutrient, available phosphates. These compounds act as a natural fertilizer in waterways, with results that can cause deterioration of the natural ecosystems.

Algae are an important part of aqueous ecosystems, decomposing organic matter and providing food to other organisms. Algae population is usually regulated by the amount of phosphates available in the water, which is generally introduced by the slow breakdown of soil and rocks. When phosphates were introduced by the disposal of detergents, however, the control on algae population was removed.

Unfortunately, the result in many waterways was an algae bloom, a rapid increase in the population of algae. As their population exploded, the algae used more and more oxygen. Other organisms that need oxygen die during a bloom, providing still more food for algae growth. Eventually, the oxygen in the water is completely depleted. No insects or fish can survive and the result is a dead area in the waterway. Algae blooms can occur in streams, lakes, and even in large areas of the ocean.

Most detergents today do not include phosphates, but it is important to keep in mind the lesson that ecosystems are extremely complex and sensitive systems of interdependent parts. Sometimes the solution to one problem creates a new, unexpected problem.

Ff **Fast Facts**

Many "dead zones" have formed in the waters off the coasts of most continents. These are areas at the bottom of the ocean where there is no oxygen. An excessive population of plankton near the surface drops large amounts of organic material to lower levels as the plankton die. Bacteria decomposing this material use all of the available oxygen, making the area uninhabitable by marine animals. Most of the dead zones form near the mouths of rivers that carry large amounts of agricultural runoff that is high in fertilizers that nourish the plankton. A dead zone of more than 8,000 square miles has formed in the Gulf of Mexico below the mouth of the Mississippi River.

Why are wetlands important?

Many environmental laws have been written in the past half century that are designed to protect wetlands. Federal and state laws define flood plains and the development that is permitted in them. Local regulations use zoning to control building near streams and marshes. What are wetlands, and why do they need protection?

According to the Clean Water Act, wetlands are areas that are inundated or saturated by surface water or groundwater at a frequency and duration sufficient to support vegetation typically adapted for life in saturated soil conditions. This definition does not require that wetlands have water on the surface all the time and, in fact, wetlands are often identified by their native plants rather than by standing water. Usually, wetlands are places where water, sediments, and dissolved minerals collect as they flow from higher elevations. Because of this collection of resources in one place, they serve as home to an amazing variety of living things.

Although the ecosystems of wetlands vary from one place to another depending on conditions such as temperature and rainfall, they generally provide homes to a wide variety of plants, insects, microbes, and amphibians which thrive in the moist environment. Birds, reptiles, and mammals are abundant because of the availability of food. Decaying plant material feeds many aquatic insects and other invertebrates, shellfish, and small fish living in surface waters.

Us **Uncommon Sense**

Wetlands are often portrayed as places overrun with mosquitoes, one reason that neighbors oppose rebuilding wetlands. Mosquitoes breed in still, warm water, and wetlands such as swamps and marshes can provide habitat for them. However, in many wetlands, water passes through in a constant flow. This does not provide breeding places for mosquitoes. In general, even wetlands with mosquito habitat tend to also provide good habitat for their predators, such as dragonflies.

Wetlands also provide valuable services to human populations. Many communities rely on groundwater for drinking supplies. The plants and soils of the wetland filter out contaminants from water as it enters the ground. They are primary sources of clean underground water. In addition, they act as natural sponges that trap rainwater and even floodwater from overflowing rivers and streams. Mats of vegetation and tree roots slow the water, releasing it gradually. Salt marshes provide the first barrier against massive sea storms. Areas in which the marshes are intact tend to suffer much less damage during hurricanes than areas where the swamps have been drained and developed.

While swamps and marshes were once thought to be wastelands, filled with dangerous animals and disease-bearing insects, environmental studies have shown that they are key elements to healthy ecosystems. Besides improving water quality and providing flood and erosion control—all very pragmatic uses from the human viewpoint—wetlands are keys sources of food and shelter for animals ranging from the smallest insects to large predators.

What is the ozone hole and how was it formed?

During the 1980s, there was a lot of concern about a hole in the ozone layer of the upper atmosphere. Chemicals in air-conditioners, aerosol cans, and certain types of fire extinguishers were replaced with other materials in order to protect the ozone layer. A truly worldwide environmental initiative was successful, leading to a slow but steady improvement. But how did the hole exist in the atmosphere in the first place?

The first step to answering the question is to describe the "ozone layer" and its usefulness to living things on Earth, including humans. Ozone is colorless gas that reacts very easily with most other substances. The sharp, acrid smell that is sometimes detectable around high voltage electrical equipment is ozone. Near the surface, it is considered to be a pollutant because it is toxic to plants and animals.

However, in the *stratosphere*, ozone is a naturally occurring substance and an important atmospheric component. The so-called "ozone layer" is not really a layer at all. Ozone normally occurs as part of the atmosphere at altitudes between 7 and 25 miles above sea level. The concentration of ozone in this region averages about 1 molecule of ozone per 10,000 molecules of nitrogen and oxygen, so it is very dilute even in the ozone layer.

Ozone in the stratosphere absorbs ultraviolet light that damages living organisms, so a change in the ozone layer is a major concern to humans. Observations of the amount of ozone in the atmosphere have shown that, beginning in the 1970s, the

> **De** | **Definition**
>
> The **stratosphere** is the second layer of the atmosphere from the surface, extending from about 6 miles to 30 miles above the surface. Because of absorption of ultraviolet radiation near the top of the stratosphere, the temperature at the top is higher than the temperature at the bottom. Large airplanes often fly in the lower part of the stratosphere because it tends to be very stable and calm.

amount of ozone in the atmosphere above Antarctica dropped substantially in the spring (September to December in the Southern Hemisphere). At its worst, the level of ozone in this "hole" dropped to about one-third of the normal level. Scientists found that the drop was caused by chlorine atoms from gases called chlorofluorocarbons (CFCs). These gases were used as propellants in aerosol sprays and refrigerants in air conditioners.

When CFCs escape into the atmosphere, they rise to the stratosphere, where they are stable for many years.

The combination of cold temperatures in the stratosphere and spring sunlight causes a chemical reaction that destroys ozone faster than it can be replaced by natural processes. A smaller ozone depletion occurs in the Arctic during the northern spring.

Although the overall decrease in the global amount of ozone was fairly small (about 4 percent), its discovery was a cause for concern. The amount of ozone destroyed was increasing from year to year, and there was a concern that holes could begin appearing elsewhere. In addition, only a small fraction of the CFC molecules in the atmosphere were destroyed, so their concentrations continued to grow.

Harmful effects of an increase in UV light could include an increase in skin cancers and cataracts in humans, reduced yields of some grain crops, and a disruption of photoplankton populations in the oceans. These concerns led to a worldwide ban on the use of CFC in 1989. Since the late 1990s, scientists have observed that destruction of ozone has decreased slightly each year.

Us Uncommon Sense

Two separate problems related to the atmosphere are often confused with one another. Global warming is not at all related to the depletion of ozone in the stratosphere. Ozone depletion is caused by the release of certain compounds that contain chlorine. It creates problems by allowing ultraviolet radiation to reach the Earth's surface but does not affect weather or climate. Global warming is caused by excess greenhouse gases, such as carbon dioxide. It is unrelated to ultraviolet radiation but traps heat so it cannot be radiated into space.

How do we know about ancient climates?

About 15,000 years ago, the climate in North America was cold and icy. In fact, much of the continent was covered by huge glaciers. From 1575 to 1585, southern Arizona suffered a severe drought, worse than any droughts in recent times. This is the type of information that environmental scientists use to analyze and predict changes in environmental conditions and climates and make predictions. How do scientists find details about weather and climate from prehistoric times?

The study of climate change over the history of Earth is called paleoclimatology. It uses a variety of methods to find out what climates have existed in the past and to chart changes in climate over time.

Tree rings provide one of the most accurate ways to find out about rainfall in the past 2,000 years or so. In a rainy year, a tree grows much faster than in a dry year, so the growth ring in its trunk is wider. Because you can get an accurate date by counting the rings, this is one of the most reliable tools for studying past climates.

Glaciers and ice sheets hold a wide range of climate information, extending as far as 800,000 years into the past. An ice core is removed by drilling straight down into the sheet. Layering in the ice is used to relate the depth of the ice to its age and the thickness of a layer gives information about annual precipitation. Scientists use pollen trapped in the ice to estimate relative plant growth for different years and ash residues to track volcanic eruptions. Small differences in the ratios of hydrogen and oxygen isotopes in the ice track changes in ocean temperature.

For data beyond what is locked up in the ice, paleoclimatologists turn to sediments and sedimentary rocks. Layer after layer of sediments in oceans, lakes, and desert floors reveal preserved animal and plant material, such as bones and pollen. Sedimentary rock layers do not give us information about year-to-year changes because the information has been compressed into layers that span long periods. They do, however, record climate in long sweeps and show periods of major change.

Ff **Fast Facts**

Although information from tree rings is generally limited to the past few thousand years, there are some cases in which they can provide much older information. Many petrified trees retain the ring information. Although it is not possible to correlate these rings to particular years, the fossils do provide information about cycles of precipitation in ancient times.

How does the greenhouse effect work?

Many factors affect the temperature at any particular place on Earth, but the average temperature depends on a balance of energy that reaches the planet and energy that radiates into space. If the planet receives more energy as sunlight than it radiates, the average temperature increases. One of the controlling factors in radiation of energy into space is a phenomenon called the greenhouse effect. What is the greenhouse effect and how does it work?

Energy from the sun arrives at Earth in the form of electromagnetic radiation, including visible light and ultraviolet radiation. About 30 percent of this energy is reflected back into space, mostly by clouds. Some energy is absorbed by the atmosphere, but most of the electromagnetic radiation reaches the surface, where it is absorbed by soil and water, which then become warmer.

Warm objects radiate electromagnetic energy as infrared radiation. The amount of energy radiated increases as a surface becomes warmer. When the Earth's surface radiates energy, some of it is absorbed by the atmosphere and some is radiated into space, effectively cooling the planet. If the amount of solar radiation coming in is balanced by the amount of infrared radiation going out, the average temperature of the planet does not change. Currently, Earth's average surface temperature is about 59°F and it is pretty stable.

Most of the molecules in the atmosphere do not absorb infrared radiation very easily. That's why it can be radiated into space. Other molecules, including water and carbon dioxide, do absorb the infrared energy. The term "greenhouse effect" refers to the action of these molecules. They absorb the energy that is radiated from the surface and trap it in the atmosphere, preventing its radiation into space. The name comes from a comparison to the glass of a greenhouse trapping heat radiated when light passes through the glass and heats the inside of the greenhouse. (Interestingly, the mechanism by which a real greenhouse traps heat is very different from that of the atmosphere, but the name has been applied and that's that.)

The greenhouse effect is important to humans and other living things. If all the infrared radiation were lost in space instead of some of it being trapped, the average temperature would be 0°F instead of 59°F.

Ff **Fast Facts**

Right now, we are in the middle of one of Earth's "ice ages." While people generally think of ice ages as periods when glaciers cover much of the land surface, that is only part of the story. The first ice age occurred about 2.5 billion years ago. The second lasted from 850 million years ago to 630 million years ago. Another occurred from about 460 to 260 million years ago. The current ice age began about 40 million years ago. During the periods between ice ages, the polar ice caps melt completely. There are cycles of warm and cool climate within an ice age that correspond to shrinking and expanding of glaciers. Right now, we are in a warm period within an ice age and the big glaciers have shrunk in Greenland and Antarctica. Earth is still much cooler now than during the truly warm periods, however. There is concern that global warming could drastically increase the rate of melting. While Earth is expected to warm as part of the long cycle, global warming may cause changes to occur more rapidly than we are able to adapt to them.

Remember that there is a balance involved. What would happen if the amount of carbon dioxide in the atmosphere were to change? Then the amount of radiation from the planet would change, altering the balance of radiation in and radiation out.

This is the concern of environmental scientists. An increased concentration of carbon dioxide in the atmosphere would be expected to absorb more infrared radiation, increasing the temperature of the planet, and the effect known as global warming.

Measurements from ice cores show that the level of carbon dioxide has varied from 180 parts per million (ppm) to 270 ppm over the past 800,000 years. Times of low concentration correspond to the cooler temperatures of the ice ages and higher concentration to warmer periods. In 1960, scientists measured the carbon dioxide level in the atmosphere as 313 ppm. It has since increased to more than 375 ppm. Most of the change is attributed to the burning of fossil fuels and deforestation.

Climatologists predict that the global average temperature will increase by several degrees over the next few decades as a result of the change in the greenhouse effect. This increase could lead to drastic changes of Earth's climates.

What causes acid rain?

Since the beginning of the Industrial Revolution, limestone and marble statues have had a shorter lifetime. Their features tend to erode and blur due to reactions of acid with the stone. This acid falls out of the sky as rain, snow, and sleet. What is the source of acid precipitation?

Normal, unpolluted rain is slightly acidic because carbon dioxide from the atmosphere reacts with water to form carbonic acid. Acid rain contains much stronger acids, nitric acid and sulfuric acid. The amount of acid in this rain is generally between 10 and 100 times that of normal rain.

Acid rain forms as a result of burning fossil fuels, such as coal and gasoline. These fossil fuels are composed primarily of carbon and hydrogen, but, because they were formed from living organisms, the fuels also contain compounds that have other elements, including sulfur and nitrogen. When these compounds react with oxygen during combustion, they form sulfur dioxide and nitrogen oxides. These oxides react with water molecules in the air to form acids.

In addition to its effects on statues and buildings, acid rain can have devastating effects on natural systems. The insects and invertebrates in streams, which form the base of the aquatic food chain, are killed in significant numbers by the acid. Even moderate acidification can prevent fish eggs from hatching and, in severe cases, it kills adult fish as well.

Us **Uncommon Sense**

The acidity of acid rain is sometimes compared to that of lemon juice or vinegar, making the point that acid rain is not really dangerous. This argument misses the point that the problem with acid rain is its effect on the environment, not whether the acid is strong enough to be harmful on human skin. Would you expect a fish to thrive if you placed it in a tank of vinegar?

In forests, especially those in high altitudes where trees are often surrounded by acid fog, plants can be damaged or killed by the acid. High acidity also makes it impossible for plants to take up some minerals from soil, stunting their growth.

The main control for acid rain is preventing the nitrogen and sulfur compounds from entering the atmosphere. Power plants that burn coal have placed systems on the smokestacks that remove sulfur dioxide before releasing combustion gases into the air. The emission-control system in a car engine is designed to prevent the release of nitrogen oxides from the exhaust.

"However fragmented the world, however intense the national rivalries, it is an inexorable fact that we become more interdependent every day. I believe that national sovereignties will shrink in the face of universal interdependence. The sea, the great unifier, is man's only hope. Now, as never before, the old phrase has a literal meaning: We are all in the same boat."

—*Jacques Cousteau (1910–1997)*

Chapter 16

Astronomy

"We should do astronomy because it is beautiful and because it is fun. We should do it because people want to know. We want to know our place in the universe and how things happen."

—John Bahcall (1934–)

Astronomy is one of the oldest sciences. People have observed the heavens for as long as people have existed. Many cultures developed accurate calendars thousands of years ago by carefully observing the motions of the sun, moon, and stars. However, it wasn't until the invention of the telescope in the seventeenth century that the modern science of astronomy truly developed. The subject matter that astronomers study is very broad: everything beyond Earth's atmosphere.

Why do we only see one side of the moon?

Every night the familiar "man in the moon" looks down over Earth. The face of the moon is unchanging. Even though the moon rotates on its axis, we always see the same side of it. The only way to map the rest of the moon is to send satellites around it with cameras. Why can't we see the rest of the moon?

The simple answer is that the moon rotates on its axis once every month and it revolves around Earth once every month, so we always see the same side of the moon. But that does not answer why this should be the case. There must be an explanation, other than coincidence, for these two cycles being the same.

The reason for the identical timing is gravity. Any two bodies in the universe exert a gravitational pull on one another. If the bodies are very large, the pull on the side that is closer is greater than the pull on the side that is farther away. This difference creates tidal effects. On Earth, these effects can be observed as water moves toward one side of the planet, making a bulge in the oceans. This bulge creates the tides. Gravity also pulls on the solid parts of Earth, causing them to bulge, but because the rock does not flow as readily as water, it is not easily detected.

Because Earth has a mass that is about 81 times the mass of the moon, Earth's tidal effect on the moon is much stronger than the moon's effect on Earth—about 20 times as large. The gravitational pull of the planet pulls the moon slightly out of shape with a bulge toward Earth. The side with the bulge is attracted toward Earth a bit more strongly than the side away from it. The result is that the moon is tidally locked to Earth so that the bulging side is always facing the planet and the period of rotation has slowed to become equal to the period of revolution around Earth.

Ff **Fast Facts**

A small lake does not have tides because tides depend on a difference in gravity, the pull of gravity itself. A lake experiences the gravitational pull of the moon, as does everything else on Earth. However, the distance between the point in the lake that is closest to the moon and the point that is farthest from the moon is very small. The distance between a point in the ocean on Earth's surface and a point on the opposite side of the planet is about 8,000 miles, so tidal effects are significant in the oceans but not lakes.

Why doesn't the same thing happen to Earth, so that it always has the same side facing the moon? Actually, the tidal effects of the moon have slowed Earth's rotation over the past 4 billion years, but the change is slow because of the differences in sizes of the two bodies. Every century, the length of a day on Earth increases by about a thousandth of a second due to the drag of tides. Given enough time, Earth would eventually become tidally locked to the moon, but the solar system is likely to be destroyed by an explosion of the sun before that time.

Do planets show phases as the moon does?

The moon does not always look the same. Sometimes it has the shape of a bright circle; other times it is a thin sliver that appears to be the edge of a circle. There is a cycle of changes in the apparent shape of the moon, called the phases of the moon. Since the planets, like the moon, are seen by reflected sunlight, do they also show phases?

First let's take a look at what causes the moon's phases. We see the moon by the sunlight that reflects from its surface. The moon does not produce any light of its own. Sunlight falls on one half of the moon's surface at all times. The specific part of the moon that is lighted changes as the moon rotates, just as the United States and India are in sunlight at different times as Earth rotates.

When the sun is on one side of Earth and the moon is on the other, the entire surface of the moon that we can see is lighted and the moon is full. If the moon is on the same side of the planet as the sun, then the entire lighted side is facing away from us, making a new moon. When the moon, Earth, and sun make a right angle, half of the part of the moon that we see is lighted and half is not lighted. This phase is called the quarter moon. In between these phases, we see different amounts of the lighted part including the crescent moon, and the gibbous moon (between quarter and full).

As Earth and the other planets revolve around the sun, the same thing happens. The planets also reflect light from the sun and the amount of the surface that is lighted, from our point of view, depends on the relative positions of the sun, the planet, and the place where we are standing. Because the planets are so far from Earth, they appear as a point of light to the naked eye. With a telescope, however, the phases become apparent. The first person to observe the phases of Venus was Galileo Galilei in 1610. His observation supported the Copernican idea that the planets revolve around the sun.

Ff **Fast Facts**

The moon does not generate its own light. Moonlight is light from the sun that is reflected in our direction. Earthlight reflects to the moon in the same way. If you look at a crescent moon on a clear night, you can often see the dark part of the moon illuminated very faintly. This illumination is reflected Earthlight.

While we can see the full range of phases of the moon, we can only observe some of the phases of the planets. For example, we cannot see the completely full phase of Mercury or Venus, the planets whose orbits are closer to the sun than Earth's orbit.

Earth can never be between the sun and the planet, the position that creates a full moon. The phases of Venus and Mercury range from gibbous to crescent. When the planet is closest to Earth, we cannot see it because all its light is reflected away from Earth, back toward the sun.

The planets that are farther from the sun than Earth also exhibit phases, but they are less noticeable. Because of the angles at which these planets can be observed, they have phases ranging from gibbous to full. The phases are less distinct than those of Venus and Mercury. The phases are difficult to observe because, as the planets move from full phase, they also become more distant and dimmer. It is not possible for a planet outside the orbit of Earth to exhibit a crescent or new phase.

What are the chances of an asteroid hitting Earth?

It's a favorite disaster movie theme: a giant asteroid bears down on Earth as people panic. The rock, as large as a small city, strikes the planet as a massive fireball, tsunamis overrun skyscrapers, and the climate is completely disrupted. Is this purely fiction or should we be worried?

Unfortunately, although the movies are fiction, the possibility of a collision between Earth and an asteroid is real. In fact, it has happened countless times in the planet's history. One theory holds that such a collision was responsible for the extinction of the dinosaurs. It would be difficult, if not impossible, for Hollywood to overstate the devastation that a really big asteroid could wreak on Earth.

Fortunately, we have a plan. NASA has begun a program to find all near Earth objects (NEOs) greater than 1 kilometer in diameter and track their courses. An NEO is a comet or asteroid that has been pushed by the gravitational attraction of a nearby planet into an orbit that brings it close to Earth. NEO discovery teams examine photographs of objects in space, looking for the things that change position from day to day.

Us | **Uncommon Sense**

When a meteor enters the atmosphere, it falls at thousands of miles an hour and heats to a bright glow. The heat does not come from friction with the air of the atmosphere. Most of the heat is actually generated by compression of the air ahead of it, in the same way that a bicycle pump is heated by compression of the air inside it.

NASA estimates that there are about 1,000 NEOs that are larger than 1 kilometer. A tracking program monitors photos of space with the goal of discovering at least 90 percent of these large NEOs by 2018. Once they are discovered, the objects are tracked to find out which of them could come close enough to Earth that there is a risk of collision. As of early 2008, the project had discovered more than 5,000 NEOs, 733 of which were larger than 1 kilometer.

It is not yet completely clear what will happen if we detect an asteroid or comet that is on a collision course with Earth. There have been a number of proposals for ways to deflect or destroy the objects, but each has its limitations. The movie solution of blasting it apart with a nuclear warhead is not practical because it would break the object into many NEOs still heading toward Earth. More practical suggestions include changing the orbit by striking the NEO with a heavy spacecraft, attaching large thrust engines, or even deflecting it with light from a powerful laser. It would take a long time to implement any of these solutions. The sooner we have notice, the better our chances of dealing with the problem.

Ff **Fast Facts**

Scientists know that asteroids and comets have hit Earth in the past. These objects are also responsible for the craters on the moon. One theory explains that the extinction of the dinosaurs 65 million years ago was caused by the impact of an asteroid several kilometers across. More recently, in what is known as the Tunguska event, an asteroid, estimated to have a diameter of several hundred meters, exploded above Siberia in 1908. The force of the explosion leveled about 80 million trees over an area of 800 square miles.

Why do we see different stars in summer than in winter?

People began looking at the stars and noticing their patterns before the beginning of recorded history. We recognize constellations in the sky as patterns of stars that stay in constant positions relative to one another. Some of the constellations can be seen year-round, but others are visible only during part of the year. Why do the constellations that we see change during the year?

Although stars do move, they are so far away that we cannot detect their motion. It is not the constellations that change during the year. Instead, it is our observation point that changes.

Ff ⟩ **Fast Facts** ⟩

Almost every culture in history has seen patterns, or constellations, in the stars. Not everyone sees the same pattern, even though they look at the same stars. For example, most people in the United States have no trouble finding the Big Dipper. For the ancient Greeks, however, those stars were part of a picture of Ursa Major, the Great Bear. In Britain, it is the Plough and in France, a Saucepan. To Hindus, it is the Seven Sages. Escaped American slaves followed the Drinking Gourd to the north.

Think about how Earth revolves around the sun. At any given place in its revolution, you cannot see the stars that are in the same direction as the sun, because sunlight in the atmosphere is too bright during the day. You only see the stars at night when the planet is between you and the sun.

So in January, you look out into space and see the group of stars that we identify as Orion, the hunter (actually Orion is visible from October to March). In July, though, Earth has moved to the other side of its orbit, so Orion is behind the sun and not visible from Earth. Now, you look up and find Libra, the scales, which was nowhere to be seen in January.

What about Ursa Major, the large bear (or the Big Dipper, if you prefer). It points toward Polaris, the North Star, all year round. Again think about Earth in its orbit. The sun never enters the sky above the North Pole or below the South Pole. It does not matter what the season, the constellations in those directions are always visible from Earth. However, the year-round stars are different in the Northern and Southern Hemispheres.

What causes the seasons?

The seasons have a huge effect on our lives. They determine what we wear, what activities we pursue in work and leisure, what we grow for food, and when we plant it. The ability to predict seasonal changes was so important to early human cultures that they built sophisticated observatories for that specific purpose. What causes the seasons to change in a regular, predictable pattern?

We experience seasons because Earth's axis is tilted by about 23.5° in relation to the plane of its orbit around the sun. As a result, the intensity of sunlight that reaches a particular place on the surface changes in an annual cycle. To see how this works, consider the most extreme cases—the North and South Poles. During the summer, the tilt of the axis means that the pole is in a direct line to the sun all day, every day. Solar energy provides heat and light round the clock. In the winter, though, the pole is always shaded from sunlight by Earth itself, so the night lasts for about six months, during which the pole receives no energy directly from the sun.

Now, what if you live in Portland (Maine or Oregon, take your pick), about halfway between the North Pole and the equator? The tilt of the axis has a major effect although much less than at the pole. In the early summer, time between sunrise and sunset is about 15½ hours, and in early winter, about 9 hours. In addition, the sunlight strikes at a more direct angle during the summer, so more energy is received during an hour of sunlight.

Ff **Fast Facts**

At the equator, seasonal changes are almost nonexistent. The apparent position of the sun changes from a little bit in the north to a little bit in the south. In Singapore, which is very close to the equator, the average high and low temperatures in January are 85°F and 72°F; in July, it's 86°F and 75°F.

Us **Uncommon Sense**

Seasonal changes are not caused by differences in the distance of Earth from the sun; however, these changes do have a small effect on temperatures. Earth is slightly closer to the sun in December than in June. As a result the Southern Hemisphere has slightly milder changes between summer and winter than the Northern Hemisphere.

Why does the moon have more craters than Earth?

If you look at the moon through a small telescope or a pair of binoculars, you can see that it is covered with craters. A detailed photograph, taken through a large telescope, shows that there are many thousands of craters on the moon, giving it a patchwork appearance. Small craters cover the floors of large craters. There are a few craters on Earth, but they are rare by comparison. Why does the moon have so many craters and Earth so few?

There are two ways that craters can form—the collapse of rock around a volcano and the impact of an object from space. Most of Earth's craters were formed by volcanoes, although there are a few known craters caused by meteors. The moon shows the opposite. There are some craters that show evidence of volcanic activity and there are some very large plains that were formed by lava flow.

Most of the moon's craters, however, were formed when its surface was struck by fast-moving objects. The energy released by these impacts cracked and melted the rock and blew it aside. The walls of the craters are made of this ejected rock. Some of the craters are surrounded by streaks of material scattered tens of miles in every direction. There are a few similar craters on Earth. Barringer Crater in Arizona looks just like a moon crater. It is a bit less than a mile wide and 650 feet deep.

Ff Fast Facts

NASA astronomers photographed the birth of a new crater on the moon on May 2, 2006. As a meteoroid struck the surface and an explosion equaled about 4 tons of explosives, a brilliant flash was recorded. Based on the brightness of the fireball and how long it lasted, astronomers calculated that the rock from space was about 10 inches in diameter and traveling at 85,000 mph. The impact left a new crater that is about 40 feet wide and 10 feet deep.

Earth is much larger than the moon, so you would expect that it would be struck by meteors more often than the moon. In fact, it is. Craters are rare on Earth, though, for several reasons. We have a protective blanket around us—the atmosphere. Most of the objects that strike the planet never make it to the surface. Intense heat, caused by the compression of air ahead of the meteor, usually causes it to disintegrate far above the surface, leaving only a momentary bright streak across the sky. Without an atmosphere to protect it, the moon's surface makes a much better target.

Even when a crater does form on Earth, it is generally a temporary geological feature. In most places, erosion by wind and water begin to erase the crater immediately. The Barringer Crater is located in the desert of Arizona, where the climate keeps these forces of erosion to a minimum. Even so, there is evidence of erosion on its walls and floor. This crater is about 50,000 years old. Some of the moon's craters have remained unchanged (except for new impacts within the crater) for about 4 billion years, nearly 100,000 times as long.

Finally, the moon has more craters because its surface is older. The surface of Earth is constantly changing due to plate tectonics, while the moon does not experience this change. As the continents are buried beneath one another and new crust forms, the materials of the surface are continually recycled, erasing any feature on them. Much of the surface of our planet is less than 200 million years old, about one eightieth of the age of the moon's surface.

How do we know the mass of Earth?

If you want to know your mass, you can step on a bathroom scale. By measuring the force of Earth's gravity on your body, the scale can tell you that your mass is 50 kilograms (110 lb.), 60 kilograms, 70 kilograms, or whatever your particular reading is. But how can you measure the mass of Earth itself?

The mass of Earth is about 6×10^{24} kilograms. That is about 6 million billion billion kilograms. How can we possibly know that value? Isaac Newton determined that the force of gravity between two objects increases as the mass of the objects increases and decreases as the square of the distance between them.

 Uncommon Sense

The period of rotation around a more massive object does not depend on density but only on total mass and the distance from the center of the object. If the sun were to shrink into a black hole the size of an orange, Earth would not spiral into it. Earth would continue orbiting at exactly the same distance. However, an object that approached very close to the tiny sun would experience an extremely high gravitational pull because its distance from the center of mass would be so much smaller.

An object stays in orbit around another object because the acceleration of gravity is exactly equal to the inertia of the orbiting object. If an object is orbiting Earth at a known distance from Earth's center, you can calculate the amount of acceleration due to gravity. If the mass of the object is much smaller than the mass of Earth, the object's mass does not matter. At a particular distance from Earth, a marble, a school bus, and a moon will all revolve around the planet at the same rate. Knowing that rate and the distance, you can calculate the mass of the planet.

We know the distance to the moon and we know its acceleration due to gravity from the period of its orbit. With those values we can measure the planet's mass.

Why don't we see an eclipse every month?

A solar eclipse occurs when the new moon passes directly between the sun and Earth, so the moon's shadow falls on the surface of the planet. A lunar eclipse occurs when Earth passes between the sun and a full moon and Earth's shadow blocks sunlight so that is does not reach the moon. Why don't we see a solar eclipse and a lunar eclipse every month?

If you illustrate the orbits of Earth and the moon on a sheet of paper, it certainly looks like there should be an eclipse every month. That is not the complete picture of what is happening, though. The lunar orbit falls in a slightly different plane than Earth's orbit around the sun. In the paper model, that means the moon is above the paper part of the time and below the paper part of the time. We only see an eclipse when the orbit crosses the plane at exactly the same time as the three bodies come into a straight line.

The picture gets even a bit more complicated. When a solar eclipse does occur, the shadow of the moon covers only a small part of Earth's surface and the eclipse lasts only a few minutes. Even when there is an eclipse, you have to be at the right place in order to see it. Total eclipses occur between two and five times per year, but any particular spot on the surface will only experience a total eclipse about every 360 years.

Ff **Fast Facts**

It is only an accident of size that we can see a total solar eclipse at all. The distances of the moon and sun from Earth are such that both bodies appear to have the same diameter. Due to tidal effects, the moon is slowing in its orbit and as a result is moving farther away from Earth by a few centimeters per year. In another billion years, the moon's apparent diameter will be too small to completely cover the sun and there will be no more total solar eclipses.

Lunar eclipses actually occur less often than solar eclipses. Because Earth's shadow is much larger than the moon, however, the eclipse lasts longer and it is visible from the entire night side of the planet. As a result, any given location can observe as many as three lunar eclipses in a year, although in some years there may be none at all. The maximum possible number of eclipses in a year is four solar and three lunar.

Why did Pluto get demoted from its status as a planet?

Most people learned in elementary school that our solar system has nine planets. The most recently discovered, Pluto, was first detected in 1930. Suddenly, in 2006, the number of planets changed. No, we didn't lose Pluto to some other star. It is still there, moving around the sun in the same orbit that it was following in 1930. Astronomers have reclassified Pluto, though, and no longer call it a planet. Why did Pluto lose its status?

Before Pluto was discovered in 1930, eight planets were known. They fell into two groups: terrestrial planets and gas giants. Terrestrial planets, including Earth, are made primarily of rocks and metals and they are small, hard, and dense. Gas giants are big balls of hydrogen, helium, and other gases. Even though the gas giants are not nearly as dense as the terrestrial planets, they are so big that their mass is much greater than Earth's mass. The giants are big and soft and they are located much farther from the sun than their smaller neighbors. Pluto was a bit of a misfit—apparently a terrestrial planet orbiting beyond the gas giants.

Over time, as more was learned about Pluto, the differences between Pluto and the rest of the gang seemed to grow. While the rest of the planets revolve pretty much in the same plane, Pluto's orbit is off that plane by about 17°. In 1978, astronomers determined that Pluto is not larger than Mercury, as had been previously believed. Its mass is actually only one twenty-fifth that of Mercury, and Pluto is only nine times as massive as the asteroid Ceres. Pluto, and its moon Charon, consist primarily of ice—frozen water—so they are not made of the same material as any other planet.

That still wasn't too much of a problem until the 1990s, when astronomers began to discover other objects in orbit around the sun in the Kuiper Belt, a region beyond Pluto. We now know of several objects in the Kuiper Belt whose size is similar to that of Pluto. The obvious question came up. Is each of these objects also a planet, even though they are so different from what has traditionally been called a planet? If so, there may be dozens or hundreds of planets and the definition might be so expanded as to be useless. If not, what about Pluto?

Us **Uncommon Sense**

Pluto is not the solar system object that is most distant from the sun. A disc of icy objects, known as the Kuiper Belt, extends about 10 times as far from the sun as the orbit of Pluto. The Oort Cloud extends even farther, one third of the way to the nearest star. It is estimated that the Oort Cloud contains as many as a trillion comets. Occasionally one of these comets is deflected by the gravity of a planet, distant star, or other object in the cloud. It may then approach the sun, creating a visual treat for observers on Earth for a few months before disappearing from sight on its way back to the Oort Cloud.

In 2006, the International Astronomy Union took up the question of what exactly makes an object a planet. One proposal was to define a planet as a body that orbits the sun and has enough mass such that gravity pulls it into a spherical shape. The problem with that was that it immediately added three planets: the asteroid Ceres, Pluto's moon Charon, and a Kuiper Belt object, Eris. Many more objects might follow, leaving the same muddled definition.

Ultimately, the astronomers classified the objects orbiting the sun in three categories:

◆ Planets are round objects that have cleared their neighborhood of smaller bodies.

◆ Dwarf planets are round objects that have not cleared their orbits and are not satellites of a larger object.

◆ Smaller solar system bodies are all the other objects that orbit the sun.

Under this definition Pluto, Charon (which is about the same size as Pluto), Ceres, and Eris are classified as dwarf planets.

Why aren't all stars the same color?

If you look at the sky on a clear night, you can see hundreds of bright stars. They don't all look the same. Brightness varies significantly from one star to another. Although they look like white points, there is some variation in color. Some, such as Betelgeuse, appear to have a red or orange tint. Others, such as Rigel, appear to be a bit more blue than most stars. If you look at a color photo taken through a telescope, you will see that stars come in a wide variety of colors. Why do stars have different colors?

Just as people are different from one another, each star has its own characteristics. They vary widely in size, with the largest stars having about 300 times as much mass as the smallest stars. They vary in brightness, too. The brightest stars emit about 100 million times as much light as the dimmest stars. However, this is not the main reason that stars appear to have different levels of brightness to an observer. That is generally due to variations in distance among the visible stars.

And the stars vary in color. The color difference is due to temperature at the surface of the star. The light from a star is caused by the same process as the light from an incandescent light bulb. Atoms near the surface of the star absorb energy and then emit that energy as light when energetic electrons return to their ground state. The color of the light that the atoms emit is related to the amount of energy that they have absorbed. A star's color is related to the temperature of the atoms near its surface.

Red stars, such as Betelgeuse, are relatively cool at about 3,000 *kelvins* (K). The yellow light of our sun is produced at a temperature of about 6,000 K. At 10,000 K, stars emit white light. The hottest stars, such as Rigel, have a surface temperature between 20,000 K and 30,000 K. Their light is blue. This does not mean that the stars emit light of only one color. Like sunlight, the light from stars of every color include the entire spectrum. Blue stars emit more light in the higher-energy blue wavelengths than in the lower-energy red. Red stars do emit some blue light but not in the same intensity as red.

De | Definition

A **kelvin** is a unit of temperature that is equal to 1°C. The zero point of the Kelvin temperature scale is absolute zero, which is –273°C. To find the temperature in kelvins, you can add 273 to the Celsius temperature.

When you look at the sky, though, almost all of the stars look white. Does this mean that most stars have a temperature around 10,000 K? No, the white appearance of stars is due to the way our eyes work. At night, when the total amount of light is low, we tend to use receptors in the eye that do not respond differently to different colors. It is as if you were to take a black and white photograph of the sky. The colors are there, but we don't have enough information to detect them.

Ff **Fast Facts**

The temperature that determines the color of a star is its surface temperature. While the temperatures range from 3,000 K to 30,000 K, the interior temperatures of stars, where nuclear fusion takes place, are much higher. The temperature at the core of the sun is about 15 million K. In a star that is about to explode, the temperature can reach as high as 100 billion K.

How do we measure the distance to stars?

It is important for astronomers to know the distance to stars because we can't know other properties such as brightness and mass without knowing the distance. But the stars are very far away, so far that we can't measure gravitational effects. How do we measure distance from our solar system to various stars?

The keys to measuring the distance to stars are knowing that Earth revolves around the sun and knowing the diameter of its orbit. To find the distance, we can take a look at a star and note its position compared to other stars around it. Six months later we take another look. The position of some stars will appear to have changed.

To see why this occurs, hold a pencil up about a foot in front of your face. Close one eye and look at a distant object behind the pencil. Now look at the pencil with the other eye. The pencil seems to move compared to the background because your eyes are several inches apart. This difference based on the point of view is called parallax.

Now go back to stars. When we look at a star in spring and then again in fall, our baseline is not a few inches, but rather about 180 million miles. Nearby stars move slightly compared to a background of very distant stars. By measuring this change we can use parallax to find the distance to the closer stars. If we use distant galaxies as our background, we can measure the distance to stars that are far from Earth.

"Space travel is utter bilge."

—Richard Woolley (1906–1986) (Statement made in 1956, one year before the launch of Sputnik.)

Chapter 17

Cosmology

"Even if there is only one possible unified theory, it is just a set of rules and equations. What is it that breathes fire into the equations and makes a universe for them to describe? The usual approach of science of constructing a mathematical model cannot answer the questions of why there should be a universe for the model to describe. Why does the universe go to all the bother of existing?"

—*Stephen Hawking (1942–)*

Cosmology is the study of the universe as a whole. The big questions of cosmology include the nature of the universe, how it began, and its ultimate fate. A basic assumption of physical cosmology is that the universe is governed by physical laws. For example, the nature of electromagnetic forces and gravity, and the basic structure of matter, are the same everywhere in the universe and they are the same now as they were in the past and the future. Like other scientists, cosmologists form hypotheses and theories and then test them with observations. However, unlike many other fields of science, the cosmologist cannot test theories by changing the conditions of the experiment. Much of the data was generated billions of years ago, although it is just becoming available to us now.

What exists beyond the universe?

Long ago, the universe was believed to consist of Earth, surrounded by the sun, the moon, the planets, and a sphere of stars. As more information has been obtained about objects in space, we know that the universe is unimaginably larger than that early model—so vast that light traveling at 186,000 miles a second takes billions of years to reach us from the most distant galaxies. Even with a universe that huge, we still ask an obvious question: what exists beyond the universe?

Because the universe is defined to be everything that exists, including matter, energy, space, and time, there cannot be anything outside the universe. Even empty space cannot surround the universe because that space would be part of the universe. On a conceptual basis, it is hard to understand how an expanding universe cannot be expanding *into* something, but a mathematical description of the universe does not have the same constraints as the human imagination.

Ff **Fast Facts**

In order to understand the universe at all, we have to make a basic assumption: the laws of the universe apply throughout the entire universe and do not vary from place to place. While it may seem obvious, it is necessary to make this assumption as part of any explanation. For example, the speed of light in a vacuum is a constant value in all of our observations. If this is not true in other parts of the universe, then any data we obtain from them is not useful in describing the universe as a whole.

There are speculations that other universes exist, perhaps even an infinite number of other universes. If there are other universes, it is possible that the natural laws that govern them are completely different from those of our universe. However, we cannot treat these ideas as scientific concepts because we have no way to obtain data from other universes, so there can be no experimental test hypotheses about their existence or the existence of anything else outside the universe that we can observe.

How do we know that the universe is expanding?

The Doppler effect, which is the change in wavelength of a wave based on the motion of its source relative to the observer (see Chapter 4), is a key element of our understanding of the expansion of the universe. What does the Doppler effect have to do with the size of the universe?

Light from stars and galaxies comes from the energy changes of electrons in atoms. Because only certain energy changes are possible, we know what wavelengths of light are emitted by energy sources such as stars. A star that is moving toward our solar system emits light in the same wavelengths as a star that is stationary relative to our solar system. However, the light that we perceive has a shorter wavelength—that is, it is shifted toward the blue end of the spectrum. A light from a star that is moving away from us shows a red shift.

The same effect occurs with galaxies that are in motion relative to our galaxy. The light from some nearby galaxies is blue-shifted, but most of the galaxies in the universe have light that is red-shifted. The interesting observation is that, with the exception of a few nearby galaxies, the farther a galaxy or other object is from the Milky Way, the greater its red shift. That means the most distant objects are moving away fastest. This phenomenon is the same in every direction we look.

If everything is moving apart faster and faster in all directions, then the universe is expanding. The observation does not mean, however, that we are at the center of the expansion, as it may seem. It doesn't matter what reference point you use, the universe is expanding in all directions from it. A common model to illustrate this concept is a balloon. If you place a number of dots on the surface of a partially inflated balloon and then continue to inflate it, what happens to the distance between dots? The distance between any two dots increases, and the farther the dots are from one another, the greater the increase. Galaxies and groups of galaxies are the dots in our universe.

Us   Uncommon Sense

Remember that the balloon model is designed to illustrate one aspect of expansion. It is not a model of the universe itself. The Big Bang is often described as an explosion. In our experience, material moves outward from the center of an explosion. This, too, is just a model to explain the beginning of the expansion of the universe. The universe, as a whole, has no center.

How do we measure the distance to faraway galaxies?

We measure the distance to stars that are relatively close to the sun by using the parallax (see Chapter 16) from observations at opposite ends of Earth's orbit. This method is useful for objects that are within about 300 *light-years* of Earth. However, the Milky Way measures about 100,000 light-years across—and most galaxies are

De **Definition**

A light-year is defined as the distance that light travels in a vacuum in one year. Keep in mind that a light-year is a unit of distance, not of time. One light year is equal to about 5.9 billion miles.

millions or billions of light-years away. How do we measure the distance to objects beyond the limit of parallax measurements?

To start, let's look at how we can measure the distance to stars in our galaxy that are farther than 300 light-years from us. Although there are many stars, they fall into a limited range of types, which can be distinguished by the spectra of the light that they emit. All of the stars of a particular type are very similar and emit about the same amount of light. The brightness of the light that we detect, however, decreases as the square of the distance to its source. If we know how bright a star really is (based on similar stars that are close to us) and how bright it appears to be, we can calculate its distance. This method works for finding the distance to the stars in our galaxy and other galaxies nearby. For galaxies that are greater than about 30 million light-years distant, the brightness of individual normal stars is not sufficient for the distance measurement.

The distance to galaxies up to about a billion light years away can be measured by a similar technique, though. Two different measures of brightness are available. First, there are certain stars, called variables, whose brightness increases and decreases in a regular cycle ranging from about 1 day to about 70 days. The brightness of these variables correlates directly to the cyclic period. The period gives a measure of actual brightness. The second measure is available when stars explode at the ends of their lifetimes. These exploding stars, known as novas and supernovas, can be billions of times as bright as a normal star but their brilliance lasts only for a few hours or days. The actual amount of light emitted depends on the type of star and can be identified by its spectrum. A nova or supernova in a galaxy is a benchmark that can be used to determine its distance.

Ff **Fast Facts**

A nova or supernova occurs when the heat of a star is not sufficient to keep the star from collapsing under its own gravitational force. When it does collapse, tremendous energy is generated in its interior, causing the star to explode. In the Milky Way, novas occur several times a year. Supernovas are rare, though, occurring on average about once a century in a galaxy such as ours. The last supernova in the Milky Way occurred in 1680.

For objects that are more distant than about a billion light years, we can measure their distance by the red shift of their light. The lengthening of light waves provides information that can be used both for determining a galaxy's distance and how fast it is moving.

How old is the universe?

Humans have been around for maybe a few hundred million years, but Earth itself is much older—about 4 billion years or more. Scientists have evidence that even the solar system is a fairly young part of the universe. How old is the universe and how do we know?

Us Uncommon Sense

We cannot know how the universe looks today by observation alone. When we look at a distant galaxy—for example, a galaxy that is a billion light-years away—we don't know how it looks now. We know how it looked a billion years ago. Also, the galaxy is not a billion light-years from Earth. Its light has taken a billion years to reach us, but the distance between us and the galaxy is increasing constantly. When the light we see now began its journey, the galaxy was less than a billion light-years away from us. Today, it is much more than a billion light-years away.

Until recently, there were two different ways to estimate the age of the universe—the ages of the oldest stars and the expansion rate. The age of a star is calculated from its mass and the rate that it generates energy. The expansion of the universe can be calculated from the red shift of distant galaxies. To determine the age of the universe, you "play it backwards" to the time when the entire universe was a single point.

Neither of these methods yields an exact result for the age and there is some disagreement between the values. If you retrace the expansion of the universe, you find an age of about nine billion years. However, the estimates for the age of the oldest stars ranges from 11 to 18 billion years. It is obvious that no stars can be older than the universe. There are several possible explanations: the calculation of the expansion of the universe is wrong or our measurements are not accurate enough; the Big Bang theory is incorrect; or there is more mass in the universe than believed. The existence of additional matter is considered a possibility because that would explain a slower expansion of the universe and there is some additional evidence that we are not able to detect everything in the universe (see dark matter and dark energy later in this chapter).

How can we detect a black hole if light can't escape it?

A black hole is a region in space with a gravitational field so strong that even light is pulled into it if it gets close enough. In fact, that is the source of its name. If light does not leave a black hole, it is invisible. How is it possible to detect a black hole in space?

The escape velocity of Earth is about 7 miles per second, which means that an object must move away at that speed to escape the pull of Earth's gravity. If the mass of the planet were doubled, the escape velocity would double. If its diameter were halved, the escape velocity would be increased by a factor of 4. That is because the pull of gravity increases with mass and decreases by the square of the distance from the *center* of the mass. On the surface of Earth, you are about 4,000 miles from its center.

A black hole that has the same mass of the sun has a diameter of less than 2 miles. This diameter corresponds to the *event horizon*, the sphere around the black hole at which the escape velocity is 186,000 miles per second, the speed of light. Because nothing can travel faster than the speed of light, nothing, including light itself, can move fast enough to escape the black hole if it is located at or inside the event horizon.

Because light cannot get past the event horizon, you cannot see a black hole directly and there can be no actual data about what happens inside the black hole. It is possible, however, to observe a black hole by its effect on other objects.

If there is a large mass in a region, its gravitational effects on other masses will reveal it. Astronomers look for black holes by looking for a large mass in a small volume that does not emit light. They believe that they have found black holes in the center of some galaxies, possibly including the Milky Way. Remember that you can determine the mass of something by the rate at which other objects orbit around it. The speeds of rotation of stars around the center of certain galaxies indicate masses in their cores that range from millions to billions of times the mass of the sun.

The second way to find a black hole is to look for the emission of energy from the material around it. If the black hole is surrounded by gas or near a star or other source of gas, atoms that approach too closely will be pulled toward the black hole. As they move faster and faster, these atoms gain a tremendous amount of energy. When they collide with one another, before reaching the event horizon, the atoms emit radiation as high-energy x-rays or gamma rays. Because this radiation is outside the event horizon, it can travel away from the black hole. These emissions are not a certain indication of a black hole because there are other possible sources for them, but they do give astronomers an idea where to look.

 Fast Facts

No one has ever directly observed a black hole. Currently, black holes are a theoretical construction based on our knowledge of physics, but their existence is consistent with a large amount of real data. Even though many astronomers believe that the core of many galaxies consists of a massive black hole, that explanation is based on a number of different observations of the behavior of matter around the core of the galaxy. The data are consistent with the theory of their existence, but scientists are seeking more evidence before stating with certainty that black holes exist. No one will ever directly observe a black hole.

Why are quasars so bright?

Quasars are the most distant objects ever observed. They are also the brightest, generating as much as 100 times the light of the Milky Way from a region that is no larger than our solar system. What are quasars and why are they so bright?

When quasars were first discovered in the 1950s, it was clear that they were much smaller than galaxies but far too bright to be stars. They were called quasars, short for quasi-stellar radio sources. More than 100,000 quasars have been detected. They are so distant that their light has traveled billions of years to reach us, so we are seeing them as they existed long ago. Most quasars appear to have been formed during the first billion years of the universe's existence.

Current theories explain that a quasar is a galaxy with a supermassive black hole in its center. Because the quasar is so distant, we do not see the stars and other matter surrounding the black hole, so its diameter appears to be very small. As matter approaches the black hole, it acquires a tremendous amount of energy due to the gravitational pull of a black hole with a mass that may be as large as 100 million times the mass of the sun. Some of this energy is generated as electromagnetic radiation, including light, radio waves, and x-rays. It is this radiation that we detect billions of years later on Earth.

There were more quasars in the early universe because they can only emit light when matter is falling into the black hole. As the gas and dust near the center of the galaxy are consumed by the black hole, it dims substantially. It is possible that most galaxies were quasars at one time and they have now become quiet.

What are dark matter and dark energy?

Recent estimates of the composition of the universe, based on its rate of expansion, observations of the galaxies, have shown that we cannot see everything. If the parts that we can see accounted for the entire universe, it would not act as it does without violating the laws of physics. Astronomers now calculate that we can only detect about 4 percent of the universe. The rest consists of matter and energy that we cannot detect, which they call dark matter and dark energy. What are dark matter and dark energy?

The existence of dark matter was proposed decades ago when scientists determined that the total mass of the stars and the dust between them is too small to generate enough gravitational force to hold the galaxies together. Also, galaxies in clusters move too rapidly for their motion to be explained by the gravity of their known matter. There must be some additional matter around us that holds things together, but which we cannot detect. This matter is called dark because it does not emit any electromagnetic radiation that we can detect.

Ff Fast Facts

Even taking into account dark matter, most of the universe is very empty. Consider that the nearest star is 4.3 light-years from the sun. There is very little matter between them and we are inside a galaxy, one of the dense clusters of matter within the universe as a whole. Regions between galaxies and between clusters of galaxies contain very little matter. By contrast, we are in a very cluttered part of the universe.

Some of this dark matter may consist of ordinary matter that we cannot see, such as very dim stars or stars that have not yet ignited. Because the amount of unseen matter is so great, astrophysicists have also proposed that there is dark matter of a different sort. This matter would not consist of protons, neutrons, electrons, or any of the known subatomic particles. They do not interact with the matter we know through the electromagnetic force, although they do interact through the gravitational force. Such matter could be all around us and we would only be able to detect it when it had sufficient concentration to be visible by its effects on gravity.

Dark energy was first proposed in the late 1990s when astronomers discovered that the rate of expansion of the universe is increasing. Because gravity works to slow the expansion, there must be a force opposing gravity to increase the rate. The nature of dark energy is not known, although a number of ideas have been proposed. Some astrophysicists believe that its source may be outside our universe. If data were found to support that theory, it would be the first evidence that our universe is not entirely alone.

What forces shape the galaxies?

A spiral galaxy, such as the Milky Way, is an amazing structure, when you consider that each graceful spiral arm is made of billions of stars. The total number of stars in our galaxy is estimated to be about 100 billion. How did these graceful structures form in the first place?

Just as there are billions of people on Earth, each with their own physical characteristics that distinguish them from one another, each galaxy has its own characteristics. Many are spirals, such as the Milky Way and the Andromeda Galaxy, which is often photographed. Many other galaxies are ellipticals, shaped like a rounded football

(American football, not soccer), often very symmetrical around an axis. Most galaxies fall into one of these types, although the variations within a type are very broad.

The spiral galaxies are all flattened disk shapes that are spinning around the center part of the galaxy. Stars wander in and out of the arms as the galaxy rotates. The arms of a spiral are not caused by the streaming of stars as the galaxy rotates. The arms themselves are density waves, just as sound is a density wave in air. The regions between the arms are not voids—they contain gases and other nonluminous material and have nearly as much matter as the arms themselves. We see the arms because the density waves create local concentrations of matter so new stars form in the arms and emit light. The cause of the waves is not clear, but it appears that they are frequently the result of the gravitational effect of nearby galaxies on the spinning cloud of stars and gases.

According to most current theories, the great majority of galaxies, including our own, formed during the first billion years or so after the Big Bang. During this period, huge clouds of gas pulled into clumps due to the effect of gravity. Frequently, a force caused the mass of material to begin spinning, making disk-shaped clouds. Within these clouds, smaller clumps pulled together to form stars. As gravity pulls the matter of the spinning cloud closer to its axis, the spin increases. Over time, the stars burn out and die and new stars form, but gravity keeps the mass of the galaxy together.

Us **Uncommon Sense**

Looking at a picture from the Hubble telescope showing hundreds of galaxies of different sizes and shapes, it is easy to think of it as a snapshot in time. Galaxies are constantly changing, though, and the light from two galaxies in a single photo carries information from an incredible range of times. While we see nearby galaxies as they appeared millions of years ago, the light we see from the most distant galaxies shows us what they looked like billions of years ago. Many of these galaxies no longer exist.

The force of gravity also causes galaxies to collide. As they come together, the stars in each galaxy exert a force on the passing stars of the other galaxy. The stars and gas of the two individuals merge into a larger galaxy. The result of the merger is a large elliptical galaxy.

Galaxies move through space as a mass of stars and other matter but, in general, they do not exist in isolation. Most galaxies appear to be part of a group of up to 50 galaxies that are bound to one another by gravity. They move together and exert forces on one another. These groups frequently form part of a cluster, which can consist of as many as a thousand galaxies.

Ff | **Fast Facts**

The total number of galaxies in the universe is not known because we cannot see them all. Some are too faint to detect with our current technology and others are hidden behind closer galaxies or clouds of dust. Using photographs from the Hubble space telescope, astronomers have estimated that there are between 250 billion and 500 billion galaxies. These estimates could increase when more powerful telescopes are trained on the distant parts of the universe.

How will the universe end?

As the study of the universe, cosmology is concerned not only with the universe as it is today, or as it existed in the past. Another key question is: In the long run, where are we going? How will the universe end?

This is a hard question to answer. We have a lot of data about the universe today (at least the little section of it that we can observe around us). We also have a record of the past in the light that is constantly reaching us from distant galaxies. It seems like we should be able to just look at the past, project through the present, and into the future. But, as with most things in science, it is not that simple.

For one thing, we still are not certain of the mass of the universe. This is a key bit of knowledge. The universe is expanding and gravity fights that expansion. If we add up all the matter in the universe that we can't see, it is not enough to pull the distant parts back together, so the universe would just keep expanding. This is where dark matter comes in. We see evidence that the gravitational force throughout the universe is greater than can be explained by what we see, so there must be matter that we don't see. Is there enough dark matter to pull the universe back together, eventually causing it to die a fiery death as an unimaginably big black hole forms?

To answer this question, astrophysicists began to measure the expansion of the universe, to see if it is slowing. That's where things became complicated again. They found that the universal expansion is not slowing, but accelerating. That discovery led to the concept of dark energy, acting as an "anti-gravity."

We do know that in billions of years, the sun and other stars will use up all their fuel and go dim. Beyond that, it is not clear. The matter in galactic clusters may, over many trillions of years, pull together into black holes, leaving an empty universe speckled with invisible holes. Another possibility is that matter will spread thinner and thinner through expansion of the universe, gradually losing energy and leaving an unimaginably cold, almost empty space of immense proportions. Still another possibility is a repeat of whatever event led to the Big Bang in the first place, starting everything all over.

We may never know whether any of these descriptions is correct. Scientists continue to collect evidence to describe the past and present of the universe. From this evidence, they expect to be able to make better predictions about its future. The concepts of dark energy, dark matter, and even the expansion of the universe are relatively new. They are all based on advances in our ability to collect data about our most distant past. As better tools and techniques are developed, the concepts will be modified based on new data. Theories about the fate of the universe will also build on the new data.

"The only reason for time is so that everything doesn't happen at once."

—Albert Einstein (1879–1955)

Part 5

Technology— Putting It into Practice

Behind each technological advance is a wealth of scientific observation. The tools and toys of our modern world are all built on the combination of science and engineering.

Some of the greatest advances of the past few decades have centered on handling digital information. Digital technology has not only changed the way we process information and the way we communicate with one another, it has extended into every aspect of our existence. The computer has extended medical technology in many ways, not the least of which is the ability to look inside a person without making a cut.

Chapter 18

Science and Technology

"There does not exist a category of science to which one can give the name applied science. There are sciences and the application of sciences, bound together as the fruit of the tree which bears it."

—*Louis Pasteur (1822–1895)*

Today, we generally think of technology in terms of the growth of machines during the Industrial Revolution—locomotives, hydroelectric power plants, manufacturing processes—or, more recently, the computer-driven advances of the past half century—personal computers, cell phones, medical imaging. In a broader sense, though, technology refers to anything humans do in order to modify nature to meet their needs. The transformation of a large stick into a club, learning to control which plants grew in a particular place, and the invention of the wheel were all momentous advances in technology.

Another way to look at technology is as the combination of science and engineering. Science is the process of learning about the natural world, and engineering is a process of solving problems. Technology is the application of scientific knowledge to design solutions.

How does a microwave oven heat food?

Until the late twentieth century, an oven always heated food by surrounding it with hot air. A gas flame, glowing electrical heating element, or even a wood fire heated a chamber in which the food was placed. Heat was transferred from the hot air into the food. When the microwave oven was introduced, it cooked in a completely different way. How does a microwave oven work?

Microwave ovens use electromagnetic radiation, related to light or radio waves, at just the right frequency to interact with water molecules. The magnetron inside the oven emits radiation at microwave frequencies in the same way that a radio tower emits radio waves. The cabinet is designed to reflect the waves back and forth inside the heating compartment until their energy is absorbed by the food.

That's why it is not a good idea to operate the microwave oven with nothing inside it. The microwave radiation is trapped and its energy increases, eventually damaging the magnetron itself.

Ff | **Fast Facts**

Microwaves pass through glass and plastic without being absorbed or heating the material. There is a metal plate inside the glass door that reflects the radiation back toward the food in the oven. Small holes in the plate allow light to pass through so that you can watch the food as it cooks. The wavelengths of the microwaves are much longer than the diameter of the holes, however, so they cannot pass through it to the window.

As the waves pass through the food in the microwave, they create cycles of electric and magnetic fields that change direction 2.45 billion times per second. The regular changes in this field cause water molecules, which are polar, to rotate. As the molecules rotate faster and faster, they bump into the molecules around them and transfer energy that causes those molecules to move faster—and become hotter.

Microwaves do not interact with most molecules, so you cannot cook very dry foods in a microwave oven unless you add water. For example, dry rice or pasta is not cooked unless some water is added. Popcorn, however, is great microwave food because of the water stored inside each kernel.

You may have heard that microwave ovens cook food from the inside toward the outside. This is not true. Energy is absorbed by water molecules in the outside inch or so and heat is transferred inward as molecules bump into one another. When you defrost food in a microwave oven, the power comes on for a few seconds and then turns off for several seconds. This allows some of the energy to transfer inwards before more heat is added.

How do light-emitting diodes work?

If you compare a flashlight that uses several light-emitting diodes to a flashlight with an incandescent bulb, you will find the LED flashlight is much brighter. Also, the bulb of the incandescent light becomes very hot over time while the LED stays cool. The battery lasts much longer as well. Why is an LED more efficient than a light bulb?

LEDs are *diodes* that emit light when a sufficient voltage is applied across them. The main part of the LED is a semiconductor chip (usually made of a combination of gallium, arsenic, and aluminum) that has two parts that are separated by a junction. One side of the junction holds a negative charge and the other side holds a positive charge. The junction is a barrier that prevents electrons from moving from the negative to the positive regions.

De Definition

A **diode** is an electronic component that allows current to pass in only one direction. Diodes are often used to convert alternating current to direct current, to convert electrical energy into electromagnetic energy, or as on-off switches in electronic circuits.

When the diode is connected to a power source, a voltage is applied across the junction, and electrons jump from the negative side to the positive side. When the negative charge (electron) and the positive charge (atom that has lost an electron) combine, the electron moves to a low-energy state and releases energy as a photon of light. As a result, electrical potential energy from the battery or other power source is converted into electromagnetic energy. The color of the light emitted is a function of the exact materials used to make the semiconductor.

Because LEDs operate without generating heat and do not have a filament to burn out, they last much longer than regular light bulbs. Until recently, LEDs were fairly expensive to build compared to other light sources. However, as less expensive semiconductors have been developed, LEDs are used for more applications.

They have been common as indicators to show when an appliance is turned on or off for some time. Other uses include tail lights for cars and trucks, flashlights, and even the flashing lights in some children's shoes. They are also used in many traffic lights, where reliability is critical and maintenance is expensive. Although they are generally more expensive to buy, LED traffic lights function for many years without a need to replace the bulbs, and they consume significantly less power. For these reasons, many municipalities have begun to replace their old traffic lights with LEDs. Once all traffic lights have been replaced, the savings in electricity will be about $200 million per year in the United States alone.

In fact, LEDs have already become so common that we often don't even notice them. Try turning off the lights in a room with a computer, router, printer, television, and several other electronic devices, and you are likely to notice that you are surrounded by LED lights.

How can a remote control turn a television on and off across a room?

Once upon a time, in the distant past, if you wanted to watch a different television program, you had to walk to the set and turn a dial. Today, however, you control your television, stereo, DVD player, and even the ceiling fan with the touch of a button on the remote. How do remote control devices work?

A television remote control (or any of the other remotes sitting on your coffee table) emits a signal from a diode that is similar to an LED. The wavelength emitted by a diode depends on the change in energy when an electron combines with a positively charged ion. Remote control devices use diodes made of silicon, in which the energy change is less than that of a light-emitting diode. Silicon diodes emit radiation in infrared wavelengths.

Because the radiation is in the infrared range, you don't see it. But a receiver in the television set does. If you place something, or someone, between the controller and the receiver, it does not work because the signal is blocked.

There is always infrared light around, so the set also needs some way to recognize the signal from the remote. The signal is a series of flashes. A digital signal can be sent by switching back and forth between two frequencies. Part of the signal is set to identify the correct device (allowing a single remote to control several pieces of equipment) and part of the signal defines the desired action.

Ff **Fast Facts**

A restaurant pager operates on a principle similar to that of the TV remote. However, the pager must be usable even if a person is not in an unobstructed line from the transmitter. To solve this problem, the pager uses radio waves, which have a much longer wavelength than infrared radiation. Radio waves are able to pass through walls and reach the pager, even in the bar.

How does a smoke alarm detect smoke?

The first sign of a fire is often smoke, small particles of partially burned material floating in the air. Smoke detectors provide early warning of danger, saving many lives every year. During the night, as people sleep, a loud alarm is much more likely to wake them up than the smell of smoke. How does the detector find smoke in the air, often before it is noticeable by any other means?

There are two types of smoke detectors commonly used in homes: photoelectric and ionization. Some smoke alarms combine the two because their sensitivities to different types of smoke vary.

A photoelectric detector uses a beam of infrared light from an LED inside a tube. A lens focuses the light into a beam which passes through the tube. A light sensor is placed at a 90° angle to the beam. Normally the light passes from one end of the tube to the other, and the sensor does not detect any of the light coming from the source. However, when particles of smoke enter the detector, they scatter the light from the beam. When some of the scattered light strikes the sensor, it sets off the alarm.

Ionization smoke detectors use about $\frac{1}{5,000}$ of a gram of a radioactive isotope, americium-241, which emits alpha radiation (a particle identical to a helium nucleus). Every second, this amount of americium-241 emits about 37 billion particles. This sounds like a lot, but it is not enough to create a health hazard.

Fast Facts

Although ionization detectors contain radioactive materials, they do not present a health hazard. The amount of radiation is very small, even though the absolute number of particles emitted seems very large. Alpha radiation is blocked by material as thin as a sheet of paper, so it cannot penetrate the plastic housing of the detector. In fact, alpha radiation is blocked by an inch or two of air. Generally, materials that emit only alpha radiation are not hazardous unless they are inhaled.

On opposite sides of the ionization chamber of the detector, there are electrically charged metal plates, positive on one side and negative on the other. A battery connected to the two plates maintains these charges. When the alpha particles strike oxygen or nitrogen molecules in the air, the collisions knock electrons away from atoms of the molecules. This makes two charged particles, a positively charged molecule and a negatively charged electron. These charged particles are attracted to the oppositely charged plates so they move toward them, causing an electric current to flow.

When smoke enters the chamber, it absorbs the alpha particles and disrupts the ionization of air. When a sensor detects a drop in the current, it sets off the alarm.

Both types of smoke detectors are very effective, but there are some differences in sensitivity. Optical detectors are particularly sensitive to very smoky fires, such as mattress fires and slowly smoldering fires. These fires produce thick smoke with large particles and have a greater tendency to scatter light. Ionization detectors respond more quickly to fast-burning flames, which produce smaller particles in greater numbers. Ionization detectors have an additional built-in security feature. If the battery begins to fail, the charge between the plates drops. This reduces the current and causes the alarm to sound a warning signal.

Fast Facts

A smoke detector only detects particles that form during combustion, so it provides no warning of carbon monoxide, a poisonous gas caused by incomplete combustion. Homes that are heated by burning natural gas, oil, wood, or other fuels should either have a separate carbon monoxide detector or a smoke detector that also integrates a carbon monoxide detector.

How does my GPS receiver know where I am?

Ancient navigators used the position of the stars to determine their locations. Later, sailors far from land calculated their location by using a compass and the sun. Today we can drive from one place to another, navigating with a GPS receiver that continually updates our current location and the route to our destination. How does this system determine locations?

In principle, global positioning systems (GPS) are fairly simple devices. Global positioning uses satellites orbiting Earth as reference points. The satellites constantly send out radio signals that are detected by receivers on the ground (or in ships or airplanes). These signals include information about the satellite's location and the exact time.

The receiver uses a clock to measure the amount of time needed for the radio signal, traveling at 186,000 miles per second, to reach it. Based on that time, it can calculate the distance to the satellite, which means it is somewhere on a sphere of points that are all the same distance from the satellite. A signal from a second satellite narrows the possible locations to a circle. Then a third signal leaves only two possible locations for the receiver. The surface of Earth is the fourth sphere, narrowing the choices to one possible point. The system includes at least 24 operating satellites at all times so that there are always three or more satellites with a line of sight to the receiver.

In practice, the process is a bit more complicated. An accurate calculation requires an accurate measurement of time with precision of a few hundred *nanoseconds* or better. This type of precision is only obtainable with an atomic clock, costing many tens of thousands of dollars. Since that is beyond the range of most purchasers of GPS receivers, it requires some mathematical tricks.

De Definition

A **nanosecond** is one billionth of a second. In the international system of units, the prefix nano- represents one billionth.

Each satellite does carry an atomic clock (and at least one backup), so the times are synchronized on the satellites. The receiver first calculates the real time from these signals, and then calculates the distance.

How accurate is a GPS measurement? Most receivers can pinpoint your location to within about 30 feet. If you can pick up more satellites, your positioning improves, perhaps to within 10 to 15 feet of your actual location. Finally, in some places, ground-based transmitters provide an additional location. Using one of these signals, in addition to the satellites, allows the receiver to calculate a position within a few feet, close enough to tell which corner of an intersection you are standing on.

Why are atomic clocks so accurate?

In the early eighteenth century, British clockmaker John Harrison won a fortune from the British Parliament for developing a clock that was accurate to within $\frac{1}{3}$ of a second per day, far better than any other clock of the time. Accurate timekeeping was an essential part of determining latitude, making Harrison's timepiece invaluable to the British navy. Accurate timekeeping is just as important today—for telecommunications, global positioning, space exploration, and even securities trading. However, fractions of a second have now become massive errors. Atomic clocks provide accuracy to billionths of a second. How are these clocks so accurate?

All clocks depend on a vibration or oscillation—a repeating motion that occurs at a predictable frequency. From ancient times many different motions have been used to record time. The rotation of Earth on its axis and its revolution around the sun were the earliest periodic motions used for keeping time. Later clocks were based on the flow of sand through a constriction, the periodic swing of a pendulum, and the rhythmic swing of a spring-powered balance wheel. Modern quartz watches use the constant vibrations of a quartz crystal exposed to an electric current to measure the passage of time.

Every cycle used to run a clock has some variability. A pendulum clock may lose a few minutes and have to be reset each day. Even an inexpensive quartz watch can gain or lose a second per day. Atomic clocks, however, measure time within about 2 nanoseconds per day, or one second in about 1.4 million years.

Atomic clocks base their timekeeping on vibrations within atoms, which are the most consistent cycles known. Most atomic clocks use cesium atoms in an excited state, that is, in which an electron is at an energy level higher than its ground state. When the electron loses energy, it emits electromagnetic energy in the microwave frequency. Because the frequency of this oscillation is exactly the same for every cesium atom, a clock can be tuned using the frequency of radiation that matches the oscillation of the electron around the nucleus of the cesium atom. In the International System of Units, one second, by definition, is the amount of time needed for a cesium-133 atom to complete 9,192,631,770 oscillations.

Uncommon Sense

Atomic clocks do not rely on nuclear decay. This is a common misconception based on terms such as "atomic energy" and "atomic bomb," in which the concept does refer to atomic decay. Like any other clock, atomic clocks use a frequency of vibration to keep track of time, in this case the vibration of charged particles in the atom.

How do power plants produce electrical energy?

The modern world runs on electricity, the flow of electrons through a material. Large copper wires carry electrons from power plants to distant factories, homes, and businesses to provide energy that powers light bulbs, giant printing presses, and everything in between. How do power plants produce an electric current?

Power plants add energy to electrons by taking advantage of the relationship between electricity and magnetism. When an electrical conductor, such as a coil of copper wire, moves inside a magnetic field, electrons move, generating an electric current. This basic principle applies to plants that burn fossil fuels, such as coal or natural gas, to nuclear power plants, to hydroelectric power plants, and to wind turbines.

Ff **Fast Facts**

Although an electric current travels very rapidly from a power plant throughout the grid, the electrons themselves do not move rapidly. As an electron gains energy, it "bumps" another electron in the wire and passes some of its kinetic energy to that electron which in turn passes it farther down the line.

Inside the generating plant of a fossil (or renewable) fuel burning plant or a nuclear plant, potential energy, stored in chemical compounds or in the atom's nucleus, is converted to mechanical energy—moving gases or liquids. The moving fluid turns the blades of a turbine, concentrating this mechanical energy into a spinning shaft. The shaft is attached to giant coils of copper wire that spin inside a powerful magnetic field. As electrons are forced to flow in the coils, current is transmitted and distributed throughout the power grid.

Hydroelectric power takes advantage of the force of gravity to move water through the turbine. A dam creates an energy gradient in the water, building a huge reservoir of potential energy between the top and bottom of the dam. As water falls, it passes through giant turbines that power the generators. On a much smaller scale, the flow of water in waves and tides has also been harnessed in some places to turn the coil of wire in a magnetic field. The blades of wind turbines have a different shape but the same function. As the wind turns the huge vanes, a shaft in the center operates the generator.

Ss **Science Says**

"We have also arranged things so that almost no one understands science and technology. This is a prescription for disaster. We might get away with it for a while, but sooner or later this combustible mixture of ignorance and power is going to blow up in our faces."

—Carl Sagan (1934–1996)

How do home computer printers work?

The demand for inexpensive printing technology grew along with the development of inexpensive computers that could be used in businesses and homes. Although many records and processes can be handled on the computer monitor, there is still a need for paper records. They can be mailed, read without a computer, posted on a bulletin board, and some people just prefer to hold a document. How do home printers work?

There are two main types of home printers, which use completely different technologies: inkjet printers and laser printers. Generally inkjet printers are less expensive to buy and more expensive to operate than laser printers. Inkjets are used more for home color printing and are more suitable for nonpaper media, such as transfers for putting images on T-shirts. Laser printers are faster and generally have a better image quality.

In an inkjet printer, ink is squirted from a nozzle as it passes over the paper. A roller moves the paper from top to bottom while the print head that holds the ink nozzles moves back and forth, shooting ink. Each print head has several hundred nozzles that are about the diameter of a human hair. They deliver tiny dots of ink to the paper. One milliliter of ink is enough to make about 100 million dots. If you look at a page printed on an inkjet printer through a magnifying class or microscope, you can see the pattern of dots.

Inkjet printers use built-in computers to control the timing and the order in which ink is fired from the nozzles on the printing head. There are two methods for firing dots from a nozzle. Some printer manufacturers use tiny heating elements to heat the ink and create a bubble. When the bubble bursts, ink hits the paper and more ink is drawn into the nozzle. There are a couple of limitations to this technology: the ink must be heat resistant and the print head must cool between bubbles, slowing the process. The second method of ink delivery uses a *piezoelectric crystal*, a crystal that flexes in an electric current. Each time a current is applied to a crystal behind the nozzle, the flexing crystal creates pressure that shoots a droplet of ink. This peizo process is faster and does not require heat-stable ink, but the print heads tend to be more expensive than thermal print heads.

De **Definition**

A **peizoelectric crystal** is a crystalline substance that generates an electric charge when pressure is applied to it. The opposite effect occurs when an electric current is applied to the crystal—pressure forms within the crystal, causing it to flex. Quartz is the most common piezoelectric crystal.

Laser printers use a rotating metal drum that is covered by a negative static electrical charge. When light strikes the drum, atoms absorb energy and the electrons are able to move. As the drum rotates, it is charged by an electrically charged wire. A computer in the printer then creates an image by shining a laser on parts of the drum, allowing the electrons to move away.

The drum is then exposed to toner, a fine powder containing plastic particles and carbon or coloring agents. The powder is given a negative charge, so it sticks to the parts of the drum that have been exposed to the laser but is repelled by the sections that still have a negative charge. The toner is then transferred to a sheet of paper. Heat and pressure melt the toner and press it into the paper.

How can irradiation preserve perishable foods?

You may sometimes see food, particularly produce and meats, with a label stating that it has been irradiated. Exposing foods to radiation extends their shelf lives and kills insects that may inhabit produce. How does the irradiation of food act as a preservative?

Irradiation exposes the food to electromagnetic radiation, in the form of gamma rays or x-rays. This radiation is similar to visible light but it has a much shorter wavelength and carries much more energy. Gamma rays are naturally produced by the nuclei of cobalt-60 atoms. The cobalt sample is stored in water, which absorbs the radiation. To treat food products, the food is placed in a chamber with thick concrete walls to absorb stray gamma rays and the radioactive cobalt is removed from the water.

Us **Uncommon Sense**

The term "radiation" has become so associated with the toxic effects of exposure that many people are concerned about exposure to irradiated foods. However, the food does not come in contact with the source of radiation, so the consumer is never exposed to any radiation from treated foods. The process is similar to the scanning of checked luggage at an airport using x-rays.

X-rays for food irradiation are produced by machines that are similar to those used for medical x-rays, but more powerful. A high-energy electron beam strikes a gold plate, producing the x-rays.

Irradiation extends the shelf life of foods by killing microorganisms on and in the product. As gamma rays pass through living cells, their energy is absorbed by molecules in the cells. Because DNA molecules are so large, they are particularly susceptible to damage by gamma rays and x-rays. When this occurs, the cells are no longer able to reproduce and the organism dies.

During irradiation, food is exposed to radiation that is several million times as strong as that of an x-ray.

Irradiation is commonly used to treat meat, especially ground meats, in order to kill bacteria and parasites. Irradiation has very little effect on viruses, however. It is also used to destroy bacteria, fungi, and insects on fruits and vegetables, which slows spoiling. Potatoes treated by irradiation do not sprout during storage because the living tissue in the growth buds of tubers is also killed by the gamma rays.

How does biometric identification work?

As security concerns become more and more important to many governments and businesses, biometric identification methods have been proposed as a way to confirm identity. Many countries now require biometric identifiers on passports or, in some cases, driver's licenses. What are biometric identifiers and how do they work?

The word biometric literally means "life measure." Biometric identifiers are measurable physical characteristics of a person that can be checked automatically. In the simplest form, information about height, weight, and eye color on a driver's license is a biometric indicator. In general, however, these are not adequate for identification because they are characteristics that can change over time, or they can be altered or masked. Generally, for security applications, the term biometric indicator is used for characteristics that do not change and cannot be easily disguised and that are unique to a person.

While automated security checks were once found only in spy movies, advances in computer and camera technologies have made them increasingly common in real life. The most common features used for identification are fingerprints and handprints, facial features, and eye scans. Each method has different advantages and disadvantages.

Facial recognition is one of the most flexible techniques because it can be used without the subject even being aware of it. This is also a source of concern about the ethics and legality of the technique. Facial recognition systems analyze photographs taken with an ordinary digital camera. A number of specific measures are taken, including distances between specific points on the face, distance between the eyes, and width of the nose. These measures are combined to a unique code. The photograph does not need to be taken from a particular angle, so cameras can acquire photographs of large numbers of people in a crowd or an airport line and compare them to a database. Ethical concerns include the collection of data on large parts of the population without their knowledge and the ability to secretly track movements.

Fingerprints were one of the first biometric measures in common use, having been used by police agencies for more than a century. Fingerprints are unique identifiers—even identical twins have different patterns. Fingerprint identification involves comparing the patterns of ridges on the fingertips with a database. Scanning and comparison are so simple and reliable that they have even been installed in business computer systems and personal digital assistants. For some applications, handprints that scan the entire hand are more useful.

The pattern of blood vessels in the retina is also unique and appears to be unchanging through life. Unlike fingerprints, retinal scans cannot be obscured by dirt or scarring, so they have become a standard method of identification for many military installations. The main drawback of retinal scans as a routine identifier is the fact that the subject must undergo a 15-second scan, keeping the eye in position.

Iris scans look at the patterns of rings, furrows, and light and dark spots in the iris, which surrounds the pupil of the eye. Like a retina scan, the iris scan provides a unique identifier that is not subject to wear or being obscured. Currently, iris scans are slightly easier to obtain than retina scans and they can be acquired through corrective lenses. Commercial systems already exist that use a computer matching program to compare iris scans to a database for positive identification. Some countries have used them as part of their immigration programs for several years without a false match. You can even buy an iris recognition system that you can attach to your computer to lock out unauthorized users for about $150.

"A good scientist is a person with original ideas. A good engineer is a person who makes a design that works with as few original ideas as possible. There are no prima donnas in engineering."

—*Freeman Dyson (1923–)*

Chapter 19

The Electronic World

"It would appear that we have reached the limits of what it is possible to achieve with computer technology, although one should be careful with such statements, as they tend to sound pretty silly in five years."

—John von Neumann (1903–1957)

Electronics is the study of the flow of electrons. Beginning with the telegraph, people have used electrons for communication. With the development of electronic components, industries rapidly developed around radio, the telephone, and later television.

Beginning in the second half of the twentieth century, electronics began to play a major role in not only communicating information but also in processing it. As engineers developed ways to make electronic components smaller and smaller, the ability of computers to perform calculations grew at an unimaginable rate. The earliest computers filled large rooms with massive vacuum tubes. Today, a credit card–sized calculator, costing less than a dollar, has more computing power.

What happens inside a computer?

When you turn on your computer, you may hear the hard drive start to spin, an occasional beep, and the soft whirr of a fan. All of these sounds are incidental to the real operation of the computer, though. The "crunching" of data takes place at an incredible rate, but completely silently. What happens inside the computer as it works?

The main operating part of a personal computer is the microprocessor, or central processing unit (CPU), which functions entirely by performing mathematical operations. This processor is about the size of a box of matches but in fact it *is* the computer. Everything else around it is there to help the processor take in data and send out results.

De **Definition**

Hardware is all of the physical components of a computer, including the CPU, hard drive, mouse, monitor, and keyboard. Software refers to the instructions that tell the computer what to do, including such things as the BIOS; operating system; and programs such as word processors, games, and e-mail programs.

When you turn on the power, it begins to "boot" (from "pulling itself up by its own bootstraps"). The first thing to come on is the Basic Input/Output System (BIOS), a small program permanently stored in the memory. The BIOS basically tells the computer that it is a computer and where to go to find out more. The BIOS tests all the *hardware* components and provides a set of instructions to the CPU about how to find and use them.

After the boot process is complete, the CPU takes over. The CPU is a huge collection of tiny electronic components. Top-of-the-line microprocessors, as I write this, have close to a billion components (most likely, more than a billion by the time you are reading it). Transistors act as switches controlling the flow of electrons, and capacitors can store or release electrons, creating on/off states that correspond to the zeros and ones of binary numbers. These are the basic units of information that the CPU processes by the billions every second.

The CPU has two functions: it manages the flow of data into and out of its millions of circuits and it carries out mathematical and logical (such as comparing two numbers) operations. Much of the flow management involves sending numbers to memory and then retrieving them when they are needed. Keeping numbers in memory when they are not actually involved in calculations frees the circuits of the CPU so they can work on more pressing calculations.

Everything that goes into the CPU, including words, pictures, and music, is coded numerically. The operating system is a program that translates all the signals coming from input devices—your keyboard, mouse, microphone, or the Internet, for example—into code that the CPU can process. It translates the output into an appropriate form for you to use—monitor, speakers, printer, and so on. Early computers used operating systems that were perhaps a few thousand lines of binary code to run simple tasks. Today's operating systems occupy large sections of a hard drive with tens of millions of lines of code that instruct the computer to perform complex operations and to control sound, graphics, and other functions of the many programs we run on our computers.

In addition to the operating system, separate programs contain instructions for particular applications. For example, a video game program contains instructions for the CPU to perform a specific type of calculation in response to the motion of your mouse and then display the result as action on the monitor.

Us **Uncommon Sense**

Although many people find that their computer runs slower after a year or two (or sometimes less), a computer does not wear out over time. The CPU and other components do not gradually become less efficient. They either work or do not work. It is possible that programs on your computer are demanding more resources than the computer has to work with. Unless you have up-to-date programs to scan for unwanted programs, the most likely cause of a slow computer is a virus or an accumulation of spy programs that use a lot of your computing ability.

How are supercomputers different from regular computers?

There are some research projects that require massive amounts of computing. For example, models of the atmosphere and oceans are so complex that climate-change researchers must consider the tiniest variables. This type of detailed analysis is typically done on a supercomputer. How do supercomputers differ from the computer on your desk?

The speed of a computer can be measured as the number of <u>fl</u>oating point <u>op</u>erations (FLOP or flop) that the computer can perform in one second. Each floating point operation is a manipulation, such as addition, subtraction, or multiplication, of two numbers represented by a string of digits.

De **Definition**

A **gigaflop** means one billion (one thousand million, in Britain) floating point operations per second. The International System of Units defines the prefix giga as one billion, tera as one trillion, peta as one quadrillion, and exa as one quintillion.

A supercomputer is one of the fastest large computers available *at the time that it is built*. That last phrase is really quite important because computer standards change very rapidly. For example, in 1988, Cray Research Corporation introduced the first supercomputer that was able to run at a speed greater than 2 *gigaflops*, an amazing rate at that time. Today, many inexpensive laptop computers process data faster than that 1988 supercomputer.

Announcements of the new fastest supercomputer come so regularly that it is impossible to say what the fastest computer is as you read these words. In 1998, the fastest speeds were 400 gigaflops. By 2006, IBM had announced its Blue Gene computer had a speed of almost 400 teraflops, a thousand times as fast. In early 2008, Sandia and Oak Ridge national laboratories announced the formation of a joint institute to develop "novel and innovative computer architectures" for exaflop computers, another 2,000-fold increase in computing speed.

So what do they do with the biggest computers out there? Supercomputers work on very complex problems that have a lot of variable parts. Imagine trying to make a model of the whole atmosphere, taking into account each little difference in heating by sunlight all around the world. Atmospheric models use a lot of data and supercomputers process the data points. Models of proteins and how they interact with other molecules in drugs and diseases could not be handled without super-speed processing. Aircraft manufacturers use them to work out the design details of huge jetliners. It always helps to avoid trial-and-error testing with real planes.

Ff **Fast Facts**

One of the motives for looking at new ways to approach supercomputing is that, using 2008 technology, the cost of electricity to run an exaflop computer could exceed $10 million per year.

How do magnetic strips on a credit card work?

If you look in your wallet, you will find that all of your credit cards and ATM cards have a long, usually black, rectangle on the back. There is a good chance that your insurance card, library card, and even your driver's license have similar strips. These are magnetic strips that hold information, such as the card number, that can be interpreted when the card is swiped through a reader. How do these strips work?

The magnetic strip is a thin layer of magnetic material. It is made by imbedding finely powdered iron in plastic and bonding the mix to the card. When the iron particles are exposed to a strong magnetic field, they become tiny magnets themselves. Information, such as the account number, is coded onto the strip by magnetic devices that are similar to the head of a tape recorder. A series of magnetic stripes are formed in the strip. These stripes can be interpreted as a zero or a one, depending on the direction of the North and South Poles.

When you pass the card through a card reader, you slide these magnets in front of coils of conducting wire. The moving magnetic fields create an electric current in the coils. The current is amplified and sent to a computer, which interprets the information.

You have to keep your magnetic cards away from strong magnetic fields. Although the tiny magnets in the strip are fairly stable and can hold their information for years, their information can be easily erased. When the iron particles in the card stripe are exposed to a strong magnetic field, their own polarization will quickly change, effectively removing the stripes and erasing the stored information.

Ff **Fast Facts**

The technique that is used to bond the magnetic strip to the card is an example of serendipity—the seemingly accidental inspiration that leads to sudden insight. According to Forrest Parry, the IBM engineer who invented the card, he had become frustrated with every attempt to find an adhesive to bond the magnetic tape to the card. His wife was ironing clothes as he told her about his problem, so she tried attaching the plastic using the heat of the iron, which was at the right temperature to bond the material without destroying it.

How do retail theft-prevention alarms work?

Retail stores often have a set of towerlike structures bracketing the entryway inside the door that set off a shoplifting alarm. If you have ever had a clerk forget to deactivate the alarm tag on a purchase you've made, you know that they really work. How can an inexpensive, disposable tag set off the alarm, and how can a clerk turn the tag off?

Electronic article surveillance (EAS) systems use tags attached to an article to determine whether the article has been authorized to leave the store. Most systems use disposable tags that do not have to be removed. Instead, some action is taken to deactivate the tag when an item is purchased so that it does not trip the alarm. Sometimes the tag is actually integrated into the product itself.

Most stores use one of three different EAS technologies: radio frequency (RF), electromagnetic (EM), or acousto-magnetic (AM) systems. The tags, which generally cost no more than a few cents, are not reusable and they are discarded by the consumer after leaving the store.

Most common in the United States are the RF systems, which have a tag containing a miniature electronic circuit and an antenna. These are usually enclosed between two layers of paper about 1½ inch square. If you separate the paper you can see the components. The circuit includes a very small coil and a capacitor. The antenna looks like a spiral of thin aluminum foil. One tower beside the door emits radio waves at a specific frequency. When the circuit detects this frequency, it absorbs the energy and emits radio waves at a different frequency, which are detected by the other tower. The tag is deactivated by passing it over a more powerful radio signal that burns out the electronic components.

In Europe, EM systems are more common. They are also frequently used in libraries because the tag can be permanently attached to a book and reactivated for reuse. The tag that is attached to the item contains a strip of material that absorbs electromagnetic energy. As in the RF system, a radio-wave signal passes between the towers.

When the wave interacts with the strip, the strip absorbs some of the energy of the wave and it becomes magnetized. Once the strip is fully magnetized, it no longer absorbs energy and the system detects the increase in the electromagnetic energy passing between the towers. The strip is paired with a thin piece of magnetic material. To activate the tag, the magnetic material is demagnetized. To deactivate the tag, it is magnetized. The presence of the nearby magnet keeps the absorbing strip magnetized all the time so that it does not send a signal to the detector.

Acousto-magnetic systems, like the other EAS systems, use a radio-wave signal to cause a response by a tag. A flexible material in an AM tag shrinks in response to a magnetic field. This material is paired with a hard magnet that exposes the strip to a magnetic field. When the tag is exposed to the radio waves from the tower, the strip shrinks and expands, resonating like a tuning fork. The resonation sends out a high-frequency sound wave that is detected by the other tower. AM systems can be used in wider doorways than the other EAS systems. The strip is deactivated by demagnetizing the hard magnet.

How does a grocery store scanner work?

They first started to appear on groceries in the 1970s—a series of black lines on a white background. Today barcode labels appear on any package you buy. Every product has a unique label that you run past a laser to bring up the price. What information do these barcodes contain, and how do they work?

Barcodes were originally designed to help grocery stores keep better track of inventory and to make checking out faster. This application has been codified in Universal Product Code (UPC) labels, which were so successful that they have expanded to encompass almost all commercial products. Each UPC label has a series of black lines, some narrow and some wide, on a white background. Beneath the lines, you can also find a series of numbers.

The numbers on the label are a unique code for the product. Every manufacturer that participates in the UPC system pays an annual fee and is assigned an identifying number—the first six digits of the code. The next five digits identify the particular product and package size. The final digit is a check code that is used to confirm that the identifying code has been read correctly.

The lines above the numbers are a code, representing the same numbers in a form that the scanner can recognize. A laser illuminates the bars and the reader detects the pattern of bars, sending the code to a computer that identifies the product. Prices are not encoded in the barcode. Instead, they are stored in the computer, available immediately on identification. This allows the store to adjust prices without relabeling the product.

Ff **Fast Facts**

The check digit on a UPC barcode is a protection against errors. Although scanners seldom misread the code, it does occasionally happen. The check digit is derived by mathematical manipulation of the other 11 digits. If the computer calculates a check number that does not match the digit on the bar, it indicates that the code was not read. This check reduces errors by 90 percent.

Product labels are not the only use for barcodes. The same technology is useful for anything that involves monitoring or tracking. Instead of using an assigned UPC code, other barcode systems use codes that are designed specifically for the purpose. Uses for barcodes include monitoring factory inventories, tracking medical samples in laboratories, following packages from pickup to delivery, and identifying hospital patients. Researchers studying honeybees have even labeled the bees with tiny barcodes to track their comings and goings.

How do radio frequency identification devices work?

In the interest of faster transactions (saving a whole 10 seconds), credit card companies have begun using contactless credit cards. You simply wave the card in front of the reader—no impressions or swiping a strip—and the transaction takes place. Actually, speed is only one consideration. Credit card issuers also believe the cards are safer because the credit card number is never recorded during the transaction. How do credit cards work without any contact with a reader?

Contactless credit cards are only one example of a technology called radio frequency identification (RFID). Instead of a magnetic strip, information is stored in a small chip on the credit card. Inside the reader, an antenna is used to send out a radio wave signal. Inside the chip, a circuit detects the signal and sends out a response containing the identification code that it stores. A receiver in the reader relays the code to a computer, which handles the rest of the transaction.

There are two types of RFID tags—passive and active. In a passive tag, the radio signal from the reader provides all the energy needed for the chip to receive and transmit information. These tags never wear out. In an active tag, the chip has a battery to provide power. Active tags can receive and transmit information for longer distances but eventually the battery will wear out, although it may last many years.

Ff Fast Facts

Contactless credit cards can be hazardous to your wealth. Researchers have found that consumers spend about 15 percent more at locations with contactless cards compared to similar locations with traditionally scanned credit cards.

The RFID tag uses information in the same way as a barcode or magnetic strip. It functions only as an identifier. Any other information is stored in a computer that is accessible to the reading device. RFID devices have several advantages over barcodes and magnetic strips: passive tags work several feet away from the reader, active devices even farther; they do not make contact with anything so they don't wear out; they can be read very quickly; and many tags can be read at the same time by a single reader.

RFID tags come in many sizes and shapes. The smallest are barely as large as a grain of sand. In bulk production, their cost, currently as low as a few cents, is expected to drop to less than a penny apiece. It is possible that UPC labels will eventually be replaced by RFID devices incorporated into every product. An entire cart of products could then be scanned simply by rolling the cart in front of a reader.

The tags are already used for a wide range of purposes. In addition to credit cards and product labeling, they have been inserted beneath the skin of pets for identification of lost animals, built into expensive musical instruments, and used to track warehouse inventories and open locked doors. RFID tags placed on the windshields of cars allow traffic to flow smoothly through toll plazas, subtracting the toll from each account as a tag identifies itself to the overhead reader.

How do solar cells convert sunlight into electrical energy?

Driving down the highway, you often see signs indicating temporary road conditions, such as construction. Frequently, the sign is accompanied by a pole topped by a flat plate, facing upward. This plate is a solar cell, which charges a battery during the day so that the sign can be lighted at night. How does a solar cell produce an electric current?

A solar cell, or photovoltaic (PV) cell, converts the energy from light into an electric current when the light interacts with atoms in the cell. Although the first photovoltaic cells were built in 1954, they were very expensive for ordinary use. Until the 1980s, these cells were used mostly to provide power to spacecraft because the ability to produce power without using heavy fuels made up for their cost. Advances in efficiency of the cells and significant cost reductions in producing materials now make them feasible for a number of applications, including providing power for buildings.

De Definition

A **semiconductor** is a solid material whose electrical conductivity is between that of a conductor and an insulator. In general, the conductivity of a semiconductor can be changed by changing its temperature or modifying it by doping with other elements. Most semiconductors used for electronics applications are made of silicon, germanium, or compounds of one of these elements.

All PV cells are made of *semiconductors*, most commonly silicon. In a silicon crystal, each atom is bonded to the atoms around it by sharing electrons. When an impurity such as phosphorus, which has more electrons than silicon, is added in small amounts (about one atom out of every million atoms), it has an electron that is not shared with another atom. Adding an impurity to the silicon is called doping. If the impurity has nonshared electrons, the doping is called n-doping (n for negative, the charge of an electron). If the silicon is doped with an element such as boron, that is called p-doping (positive). Boron has fewer electrons than silicon and it provides a space where an electron could be shared by two atoms.

If a layer of n-doped silicon and a layer of p-doped silicon are placed together, some of the electrons move from the n side to the p side. However, as the electrons fill spaces a boundary forms. The two semiconductors become a diode. Electrons will only move past the boundary if they have a certain amount of energy. That's where light comes in. When light strikes the n-doped semiconductor, electrons of the doping atoms absorb energy and begin to move across the boundary to the p-doped semiconductor.

This cannot continue for very long, though, because a negative charge will build on the p side and repel electrons trying to cross the boundary. However, if you connect the two sides of the diode with an external wire, the energetic electrons can continue moving through the wire and back to the (now positively charged) n side.

An electric current is simply a flow of electrons in a conductor, so you have created an electric current. The energy of the moving electrons can be used to turn on a light, run a motor, or charge a battery for use later. This current will continue to flow as long as light adds energy to the electrons on the n side of the cell. The amount of electric current produced by the PV cell is directly proportional to the amount of light that strikes it.

 Uncommon Sense

Articles about solar cells sometimes state that they "convert photons of light into electrons." This is a misconception. The photon of light is absorbed by an atom and the added energy can cause an electron to leave the atom. The electron already existed, however, so no electron was created by the process. The light simply added energy to the electron.

How do digital cameras take pictures?

How often do you take film to the drugstore to be developed? If you are like most people, you no longer use film cameras, having adopted a digital camera instead. Photography is a field that has been drastically affected by the development of inexpensive computers. How do digital cameras differ from film cameras?

In general, the two types of cameras work the same way: light reflected from objects in front of the camera is captured through a lens and focused on a detector that records an image of the objects. The difference lies in the type of detection used to convert the light into a recorded image. On film, fine grains of light-sensitive chemicals change when they are exposed to light. Once the film is developed, these changes are permanent and the film cannot be reused.

In a digital camera, light strikes an array of electronic devices, called charge coupled devices (CCD), which are sensitive to light. When light strikes a CCD, electrons flow and cause a charge to accumulate. A computer then records the charge on each CCD element digitally and combines them to form a picture. The response of each CCD represents one tiny dot, called a pixel, in the final picture.

However, a CCD does not respond to color in the same way as the chemicals in film. Because each CCD is either on or off and it responds to any light that strikes it, camera makers have to trick them. Color filters are placed above each electronic element. The filter allows only one color of light to pass, so each pixel responds only to its own color. The computer assigns that color to the pixel representing that CCD. Just as the image on a computer monitor is made of an array of tiny colored dots, a digital photo consists of an array of colored dots. Every CCD response makes up one pixel of the final photograph.

Early digital cameras were not able to produce images whose quality equaled that of film cameras. However, as the size of the CCDs becomes smaller and their efficiency is improved, camera manufacturers are able to add more pixels as well as improve their performance. The quality of a digital photograph depends on the number of pixels, the type of CCD used, and the quality of the camera lenses that focus the image on the CCD array. Most digital cameras today are able to make a print that has a resolution that is as good as a 4"×5" print from a film camera. The top-of-the-line digital cameras are approaching the resolution of the best film cameras, although many professional photographers prefer to use film for some types of photographs.

Ff **Fast Facts**

The resolution of a digital camera is measured in the number of pixels that it has, which corresponds to the number of CCD elements. Cameras with a resolution of fewer than 250,000 pixels are essentially toys. Between 250,000 pixels and 1,000,000 pixels (1 megapixel), the pictures are suitable for snapshots or website photos. For printing photographs, you should have about 1 megapixel for 3"×5" photos, 2 megapixels for good 4"×5" photos, and 4 or more megapixels for larger prints. Professional photographers use cameras with very high-quality lens systems and about 11 megapixels to produce photographs whose resolution is similar to that of large-format film.

Why are there places that a cell phone does not work?

Cell phones are generally a reliable means of communication—so reliable that many people have abandoned land lines and use their cell phone exclusively. Even so, there are always those annoying moments when the call disappears and occasional places where there is no signal. Why do dead spots exist where a cell phone does not work?

First, let's take a look at how cell phones work. Mobile wireless phones have been around since the 1940s. The early systems used a few towers, generally one per metropolitan area, which sent and received signals as radio waves. The tower communicated to the mobile phones by radio and then sent the transmission on using regular land lines. By broadcasting strong signals at a number of different frequencies, these towers could handle tens or hundreds of calls at a time. When calls are expensive and mobile phones are rare, this works, but the number of available frequencies is limited, so there is no room for increased demand.

The solution to the problem of limited availability was the cell concept. In a cellular system, there are many towers within an area that once had only a single tower. Each tower covers a much smaller area, known as a cell. When you make a call on a cell phone, you communicate on a frequency that may be in use by callers throughout a city, although you are the only caller using it in a particular cell.

Ff Fast Facts

A satellite system, the Iridium network, was started in 1998 to provide cell phone coverage worldwide without any dead zones. The plan was to have 77 satellites covering the whole earth. However, the satellite network was extremely expensive and ground-based networks grew rapidly, so the system was unable to compete in price. The company filed for bankruptcy less than 10 months after launching the system. An investment that cost about $6 billion was sold for $25 million. Today the system functions with 66 satellites, primarily serving the U.S. Department of Defense, large corporations, and very remote scientific research stations.

One big advantage of a cellular phone system comes from computers that can monitor the system and track your call as you move from one cell to another. Usually, you are transferred from one tower to another with no noticeable change. Computerized systems can also handle digital signals which are much more efficient for transferring data. As the number of towers has increased, the number of calls each tower can handle has also increased to the point where it seems that everyone is always on the phone.

Another advantage of closely spaced cell towers is that cell phones need a much less powerful signal. Some older mobile phones had a battery pack the size and weight of a brick. Today you can carry a phone in your shirt pocket and still have room for a business card holder. The difference is a combination of miniaturization of electronic components and the reduced need for battery power to send a strong signal.

There is a downside to a lower power output, though. Early cell phones broadcast a strong signal that could be detected by a tower 50 miles or more distant. Today's cell phones are designed to operate within a few miles of a tower. In a rural area, you may be too far from a network tower and therefore out of the coverage area. Also, the weaker signals are more likely to be blocked by hills, buildings, or even dense foliage. Dead zones can therefore occur even in places where there should be coverage. Wireless service providers are continually adding towers and antennas, as well as improving transmission systems.

"The most profound technologies are those that disappear: they weave themselves into the fabric of everyday life until they are indistinguishable from it."

—Mark Weiser (1952–1999)

Chapter 20

Medical Technology—Looking Inside the Body

"We must not forget that when radium was discovered no one knew that it would prove useful in hospitals."

—Marie Curie (1867–1934)

Historically, the inner workings of the human body have been a mystery. Organs and tissues work together, hidden beneath layers of skin and muscle. Doctors today, however, can see and hear beneath the surface of the body, watching the brain function and hearing blood flow through arteries.

Medical technology has combined science and engineering to provide doctors with tools that could not have been imagined a century ago, or, in some cases, 10 years ago. Imaging technologies provide tools to look at specific organs and even trace the activity of individual cells in response to a stimulus or a drug. New surgical techniques send a camera inside a vein so that a surgeon can operate on the heart through a tiny incision in the arm or leg. Some procedures have even been performed by specialists, using robotic tools and an Internet connection, on a patient thousands of miles away.

A large part of the advance in medical technology comes directly from advances in computer technology. As computers become faster and more powerful, they are able to process enormous quantities of data. This allows doctors to obtain images that would otherwise be impossible and to control movements of instruments within a fraction of a millimeter.

How does arthroscopic surgery work?

Severely torn knee ligaments are painful injuries that heal slowly. These tears often afflict athletes, runners, and other active people whose activities are hard on knee joints. Surgery to repair a torn ligament, if the knee is cut open, requires months of recovery and physical therapy. Most ligament repairs today, though, use arthroscopic surgery techniques. Recovery time is normally 4 to 6 weeks. Some athletes are back to their sport in a month. How does arthroscopic surgery work?

To perform arthroscopic knee surgery, the surgeon makes three small incisions, less than $\frac{1}{2}$ inch long, into the knee. An *endoscope* is inserted into one of the cuts and remote-controlled tools into the others. Images pass through optical fibers and are projected onto a monitor so that the doctor can see exactly what is going on inside the joint.

De Definition

An **endoscope** is a device that uses fiber optics and powerful lenses to view inside the body. It has two fiber-optic systems, one to carry bright light from an outside source to the site and the other the carry images from the tip of the endoscope to a viewing lens or monitor. The endoscope may also have tubes through which small instruments can be manipulated. Endoscopes are used for diagnostic purposes, such as colonoscopy, and as surgical tools.

Miniature tools, including scissors, forceps, suction tubes, and brushes, are guided to the site of the damage from the other incisions. Orthopedic surgeons even employ tiny power tools to grind away excess bone and other hard tissue. Debris is carried away through a suction tube. Because of the magnification and the size of the tools, it is possible to do very precise work at the site of the injury.

Patients recover quickly from arthroscopic knee repair because there is very little damage to muscles, tendons, and other tissues in the joint. This type of surgery is normally performed as an outpatient procedure and the patient can go home the same day. In addition to knee surgery, arthroscopic surgery is used in the shoulder, elbow, wrist, ankle, and hip.

How are materials chosen for replacement body parts?

When a bone is badly broken, it must be held together, sometimes permanently, with metal screws and metal plates to support it during its healing. Damaged valves in the human heart can be replaced by mechanical valves made of steel and plastic. The function of badly worn knee joints can be improved with plastic pads that replace missing cartilage. All of these treatments rely on long-lasting materials that can be used inside the human body. What types of materials can be implanted inside the body?

The human body is not a very friendly place for materials. It is warm, moist, salty, and filled with bacteria as well as cells whose specific function is to destroy anything that doesn't belong there. Researchers have to carefully investigate the chemical and physical properties of any materials that are to be placed inside the body.

The earliest successful *implants* were bone plates, first used in the early 1900s to hold bones in place while they healed. These plates used a form of stainless steel made with vanadium, which was specifically designed for use inside the human body. Today, many metal implants are built of titanium, a metal which is strong, light-weight, nonmagnetic, and unreactive.

> **De** | **Definition**
>
> An **implant** is a medical device that replaces a biological structure. Implants include such things as artificial joints, heart pacemakers, stents that keep arteries open, and even devices that deliver drugs at a controlled rate.

Implanted materials cannot all be made of metal. For example, in knee replacement, some sort of cushion, usually one or more plastic pads, must be inserted between the metal or bone parts to keep them from rubbing together. The material for these pads must be chosen carefully. If the plastic wears out, forming tiny particles, the body's immune system can identify the particles as foreign objects. White blood cells are mobilized, causing the surrounding area to become inflamed. This inflammation can lead to autoimmune reactions that destroy bone cells around the implant.

Many implants are now made with ceramic parts that have better wear characteristics than plastic. Ceramics also tend to form bonds with bone, making them much stronger and more durable. Over time, bones grow new tissue that extends into the ceramic material.

Much of the current research into implant materials focuses on ways to make the material work together with living tissue, and even become part of it. Porous materials can provide surfaces into which tissues can grow and bond with the implant. Another approach that may allow the body to accept materials is coating the material with living cells. If these cell coatings are grown from the patient's own cells, then the body recognizes the implant as part of the body rather than as an invader.

How do x-rays show a broken bone?

A staple scene in cartoons is a character walking in front of an x-ray machine. You see the character walk past a screen and suddenly you're viewing his skeleton. Although doctors use a short snapshot instead of continuously watching a screen, that is pretty much what happens when an x-ray image is taken. How do x-rays work?

X-rays are a form of electromagnetic radiation just like visible light, and like visible light, they pass through some materials and are absorbed by others. If you place a strip of black tape on a window, it makes a shadow when sunlight passes through the window. The tape absorbs all of the light while the glass lets it pass through.

Ff **Fast Facts**

The discovery of x-rays and their ability to make pictures of bones was an accident. In 1895, Wilhelm Roentgen noticed that a fluorescent screen in his lab began to glow while he was experimenting with electron beams. When he put his hand in front of the screen, he saw an image of his bones.

When atoms absorb light, the addition of energy causes the electrons to move into a higher energy state. This can only happen when the energy of the light matches the change in energy of the electron. Electrons can only gain specific amounts of energy, and these specific amounts differ for different elements. Light consists of photons, packets of specific amounts of energy. When the energy of a photon exactly matches the amount of energy needed to move an electron to a higher energy level, the atom absorbs the photon.

The atoms that make up your skin and other soft tissues absorb the energy of visible light photons. That is why you cannot see through your hand. The photons of x-rays, however, have much more energy than the photons of light. They pass right through your skin, muscle, and blood without being absorbed. These tissues are transparent to x-rays in the same way that a glass window is transparent to visible light.

Your bones, however, are made of different elements than are your soft tissues. Bones (and teeth) are primarily made up of calcium and phosphorus. Electrons in these elements can gain energy in amounts that match the energy of an x-ray photon, so bones absorb the x-ray photons, just as black tape absorbs photons of visible light.

In an x-ray machine, the radiation comes from exciting atoms with an electric current, similar to the way a light bulb produces radiation in the visible part of the spectrum. The radiation passes through an opening and travels to a piece of film that absorbs x-ray photons. If something that absorbs these photons, such as a bone in your arm, is placed between the source and the film, a shadow of the object appears on the film. This shadow is the familiar x-ray image that your doctor uses to determine whether the bone is damaged.

Us **Uncommon Sense**

Some people avoid getting dental x-rays due to fear of harm from the x-rays themselves. While x-rays can cause damage to cells in large amounts, the amount of radiation exposure during a dental x-ray is very small—many times less than annual natural exposure to radiation from space. The risk from undetected dental problems is greater. When x-rays are taken of women who are pregnant, a lead-lined blanket is usually used to avoid exposing her fetus to any radiation. X-rays that strike lead are completely absorbed.

How do CAT scans make images of your body?

An x-ray image is like a shadow of an object placed between the source of the radiation and the film. It is useful for detecting damage on a tooth or showing whether a broken bone has been set straight. But if you need to see the back side of a tooth or an organ that is hidden behind bone, you will need a CAT (or CT) scan, which is spoken as "cat scan." How can a CAT scan provide images of hidden parts of the body?

If you read its name, computerized (axial) tomography (CT or CAT) sounds pretty complicated. However, you already know what computerized means; tomography is an image that is collected in sections, and axial just means all around, or from every direction. And a CAT scan is just what it sounds like: a series of images, taken from different directions, and then combined by a computer.

In a CAT scan, an x-ray beam moves in a full circle around the part of the body to be scanned. It takes hundreds of x-ray images from many different angles. For a full body scan, the scanner moves from head to toe, taking a set of images, each of which is like a single cross-section of the person. Then the computer comes in. It would be impossible for a person to look at all these images and interpret them as a three-dimensional picture. For a computer, though, that sort of thing is a piece of cake. The computer processes the entire sequence of images and builds a composite picture of the patient's insides.

Although soft tissue, including many organs, does not normally show up in an x-ray image, CAT scans are very important tools for diagnosing problems with organs. Prior to the scan, the patient is given a contrast agent, or contrast dye. This is a soluble material that absorbs x-rays. The dye is designed to concentrate in the organ to be looked at, or to remain in the blood in order to study arteries or veins.

How does an MRI scanner make an image?

Images obtained from x-rays and CAT scans provide a lot of information about bones and other hard parts of the body. With a contrast dye, CAT scans can also show doctors what is going on in internal organs. However, for a detailed picture of the soft tissues and organs of the body, an MRI scan produces pictures that are far better than scans based on x-rays. How do MRI scanners produce images of soft tissue?

Compared to x-ray technology, magnetic resonance imaging (MRI) is a fairly new way to see inside your body. It is a noninvasive technique that does not expose the patient to any potentially harmful radiation and does not require any probes inside the body. By adjusting the position of the patient and the settings of the instrument, doctors can obtain a detailed picture of internal organs, blood vessels, bones, and other tissues.

An MRI machine uses a combination of powerful magnetic fields and radio waves to produce its images. When a hydrogen atom is near a strong magnet, its nucleus lines up with the magnetic field. Most molecules in the human body contain hydrogen atoms, so they are sensitive to MRI. The atoms on different molecules respond slightly differently from one another, increasing the ability to tune, or focus, the instrument.

The MRI machine has a giant magnet that creates a very strong magnetic field that runs the length of a large tube. As the patient lies inside the tube, all of the hydrogen atoms align themselves with the field. When energy is added in the form of radio waves, some of the atoms absorb energy and move out of alignment with the field. When the radio-wave source is turned off, these atoms move back into place and release energy as radio waves. This energy is detected and a computer converts the data into a digital signal that can be converted into a picture. By properly tuning the instrument, the doctor can look at very small locations in a particular organ or tissue.

Unlike x-rays and CAT scans, MRI does not require ingestion of a contrast agent to make an image of an organ, such as the liver or a blood vessel. Tuning magnets in the instrument allow it to look at specific places. Also, soft tissues generally contain a lot of water. Two of the three atoms in a water molecule are hydrogen, so MRI images are especially sensitive to water. Diseases such as cancer and inflammation, which cause fluid to accumulate, are particularly easy to study by MRI compared to other techniques.

Ff **Fast Facts**

MRI demonstrates how quickly technology can advance when the basic science is well known and understood. The first MRI exam of a human occurred in 1977, taking almost five hours to complete. Due to advances in computer technology, magnet technology, and electronics, today's MRI scanners obtain much more information in a few seconds.

MRI is a safe and noninvasive way to look into the body, although some people become claustrophobic when they are inside the tube. The greatest risk of MRI comes from the effects of its very strong magnetic field on metal objects. MRI technicians must be certain that neither they nor the patient bring objects such as keys, stethoscopes, or paperclips into the area. These normally harmless items can become deadly projectiles when they are close to the powerful MRI magnet. Buckets, oxygen tanks, and even a police officer's sidearm have been known to fly across a room magnetized by a MRI magnet.

How does a PET scan differ from other imaging techniques?

One of the main goals of imaging technology is to learn what is happening inside a person's body without needing to cut into the body. While MRI and ultrasound look into the body, and x-rays and CAT scans look through the body, nuclear imaging uses information that starts inside. Positron emission tomography (PET) is one example of a nuclear imaging technique. How does a PET scanner make an image of an internal organ?

PET produces images by looking at radiation given off by radioactive atoms. Radioactive atoms of many elements can be made in large particle accelerators, which cause atoms to smash together at nearly the speed of light. When these radioactive atoms break apart, they emit radiation that can be easily detected.

For PET scans, radioactive carbon, fluorine, oxygen, or nitrogen atoms are used to make compounds that are injected into the patient. The patient is placed in a tubular scanner that converts the radiation that is emitted into electronic signals. The scanner moves back and forth around the area of interest and then a computer assembles the data into a three-dimensional image. The image shows where the radioactive material has accumulated.

PET scans are often used to detect cancer. The radioactive atoms are used to make sugar molecules that are injected into the body. The radiation will be higher in regions of the body that metabolize the sugar faster. Metabolic rates are higher in cancer cells than in the normal cells around them, so tumors show up very well in PET scans.

Other uses of PET include tracing the flow of blood through the circulatory system to detect areas of impaired flow and detected areas of the brain that are not functioning properly. PET is also used in research studies of the normal function of the brain. The scan shows which parts of the brain are most active as the research subject performs a particular task.

PET uses radioactive atoms that decay very quickly to minimize the patient's exposure to radiation. However, this does limit the capabilities of the technique. Because the radioactive material must be used within a few days, or sometimes a few hours, from the time they are made, most PET facilities are located near a particle accelerator.

How do defibrillators save a person having a heart attack?

On just about every episode of a television hospital drama, someone goes into cardiac arrest. A team rushes in with a cart full of electronic equipment, places paddles on the patient's chest, cries "Clear," and administers a shock that restarts the heart. How can an electric shock cause a heart to start beating?

A heart attack occurs when blood flow to the heart, or to a section of it, becomes blocked. There are two severe problems linked to a heart attack that can be treated by defibrillation, or administering an electric shock to the chest. Heart failure occurs when the heart cannot pump enough blood and, sometimes, completely stops beating. Arrhythmia, or irregular heartbeat, is any change from the normal pumping sequence of the heart.

One particular type of arrhythmia, ventricular tachycardia, occurs when the ventricles (lower chambers) of the heart beat very rapidly. When this occurs, the heart can start to quiver without actually pumping blood. Tissues throughout the body die within a few minutes if blood flow stops, so quick action is necessary to start the blood pumping. In many cases, an electric shock can start the flow again.

When the heart is functioning normally, cells in the heart, called pacemaker cells, send chemical signals that are converted to an electrical impulse. This electrical impulse is carried to the heart muscle by nerves, signaling the heart to contract and pump blood. When these signals become uncontrolled, the heart does not beat with its normal rhythm.

A defibrillator is a device that delivers electrical energy near the heart. A sudden jolt of electricity causes all the heart muscles to contract at once. This frequently ends the arrhythmia and the heart resumes its normal pace. If the heart has stopped beating, this sudden contraction can push it back into motion.

The paddles used in hospitals, and prominent on television shows, represent only one type of defibrillator. Many people have implanted artificial pacemakers. These battery-operated defibrillators administer a shock to the heart at each beat, taking over for the natural pacemaker cells.

Us **Uncommon Sense**

If you know about defibrillators only from television, you may assume that they always restart the heart. Unfortunately, this is not the case. A study at hundreds of hospitals found that, if the shock was administered within two minutes of the time that the heart stopped beating, a patient had a 39 percent chance of survival. After five minutes, the survival rate dropped to 15 percent. Still, these numbers are better than the 0 percent rate without treatment. If a heart attack is recognized and treatment is started before the heart stops beating, survival is much more likely.

In the past decade, the technology of defibrillation has advanced significantly. Defibrillators have been part of normal ambulance equipment for a long time. Trained operators have saved many lives with them. Recent defibrillators, however, have dispensed with the trained operator. They have a computer that is programmed to analyze the heartbeat and determine whether a shock is needed. This capability means that they can be used by someone with little or no training since they will not send a shock if the heartbeat is normal. These defibrillators have now shown up in airplanes, police cars, and at senior centers as essential equipment.

How does a sonogram make a picture of a fetus?

When was your baby's first picture taken? Fifty years ago, it might have been a photograph taken in the hospital shortly after birth. Today, that first "picture" might occur seven months earlier. Ultrasound technology has allowed obstetricians to make images during the early stage of pregnancy to monitor development and detect or avoid problems during the pregnancy. How can an image be made with sound?

A *sonogram* uses ultrasound—very high frequency sound waves that cannot be detected by the human ear. Images are formed by recording echoes of the sound waves from an object, such as an organ or fetus, inside a person's body. This technique is similar to sonar detection used by submarines to safely navigate around obstructions deep underwater. Every second, millions of pulses of sound enter the body. When the sound waves strike a boundary between two different tissues, an echo returns to the detector. A computer calculates the distance the sound has traveled and uses that information to make an image.

De Definition

A **sonogram** is an image formed by bouncing sound waves off an organ or tissue and recording its echo.

Sonograms are widely used in obstetrics because they do not use any ionizing radiation, which could be harmful to the rapidly growing fetus. They provide information with sufficient detail to determine sex and facial features very early in the pregnancy. Because ultrasound images are available in real time, they are also useful for guiding other, more invasive procedures.

A fairly recent use of ultrasound in medicine uses the Doppler effect, a change in frequency due to the motion of the reflecting object relative to the source of the sound. Doppler ultrasound can be used to measure the flow of blood through the heart and inside arteries.

"Advances in medicine and the possibilities of human happiness created by the relief of suffering are a great embarrassment to those determined to think nothing but evil of science."

—Peter Medawar (1915–1987)

Part 6

Science: Past to Future

In this part, we take a look at some of the big ideas that have grown through the history of science and a few of the people who have had a hand in building them.

As the list of answered questions grows, the list of unanswered questions swells. We will also take a look at some of the questions driving scientists today. These big questions create a cascade of smaller questions that will be answered one at a time. In the end, the answers will come, leading to new scientific knowledge and the technologies of the future.

Chapter 21

A Few Big Ideas

"The study of an idea is, of necessity, the story of many things. Ideas, like large rivers, never have just one source. Just as the water of a river near its mouth, in its final form, is composed largely of many tributaries, so an idea, in its final form, is composed largely of later additions."

—*Willy Ley (1906–1969)*

When the Greek scientist Archimedes screamed "Eureka! I have found it!" he was running through the streets of Syracuse naked. His exuberance came about when he realized how to measure whether King Heiros's crown was pure gold or a mix of gold and silver. He'd conceived of a test while in the bath, after observing that his body displaced water. Since gold is denser than silver, he reasoned, a mixed crown would displace less water, and that's what happened when he tested the method that became known as Archimedes' Principle. Although the idea was new, it was not a fluke. The big ideas that guide science come from a combination of careful observation and the thought processes that interpret the observation.

The history of science is filled with big ideas that become the starting point of a series of research projects that may span centuries. Each researcher adds another observation, another hypothesis, a new test, or even a new theory about the idea. The ideas in this chapter are not necessarily more important or encompassing than other big ideas. They are presented as examples of how science grows from an initial idea to one or more major theories.

Atoms

At the heart of our understanding of matter is the concept of the atom—the smallest particle of an element that has the properties of the element. The word "atom" is of Greek derivation, from *atomos*, meaning "indivisible," or more specifically *a* meaning "not" and *tomos* meaning "cuttable."

One ancient Greek view of the atom was proposed by the philosopher Democritus. According to Democritus, any substance could be cut in half, and then halved again and again, until the point where no further division was possible. Since these atoms were thus "uncuttable," they held the basic nature of each substance. An atom of gold, for example, when revealed in its essence, would be dense, malleable, and smooth to the touch, like anything made of pure gold. The atom might also have edges, so that it would fit together with like atoms and thus form a solid metal. Atoms of air, in contrast, would be spaced widely apart and rather insubstantial in construction.

Democritus believed that everything in nature formed by random collisions of unseen atoms. There were atoms, and where there were no atoms, there was a void. The idea was not universally accepted among Greek philosophers. Aristotle, for example, thought that everything observable was composed of four elements: Earth, Air, Fire, and Water. These things had no atoms but were instead continuous. If the void existed, it would violate basic physical principles.

Ff **Fast Facts**

The "uncuttable" atom that the Greeks could not measure is now known to be about 300 millionths of an inch across. To put this in human and metric terms, the width of a human hair is about one-tenth of a millimeter, and the width of an atom is a millionth of the width of a hair.

The Greek concept of the atom was not, however, a theory in the way that we use the term today. The tools and techniques to build experiments to test the idea did not exist. The ideas of Democritus and Aristotle competed, but they were essentially thought exercises. There were no scientific tests to support one over the other.

Dalton's Model of the Atom

In 1801, John Dalton presented a series of "Experimental Essays" to the Manchester Literary and Philosophical Society, of which he was a secretary. The papers were about the nature of gases. While studying properties of a wide range of materials, he determined that the ratio of elements in a compound is always the same, no matter how much of the material was analyzed. Dalton concluded that these fixed chemical combinations resulted from the interactions of atoms of various elements, each of which had a specific weight. He published a table of atomic weights, listing carbon, hydrogen, oxygen, nitrogen, phosphorus, and sulphur, with the lightest atom, hydrogen, given the weight of 1.

Dalton proposed that all elements are composed of a basic unit, which he called an atom, from the Greek concept of a smallest unit of matter. All of the atoms of a given element are identical and had the same mass. When atoms of particular elements combined in specific ratios, compounds are formed. Chemical reactions involve the rearrangement of combinations of atoms. Although our understanding of atoms differs significantly from Dalton's, his theory of atoms provides the basic concepts of modern atomic theory.

Compared to the ancient Greeks, Dalton had a couple of advantages in developing his idea. The first was the development of the scientific method, which provided a structure to his investigations and a way to support his theory. The second was that a number of elements had been identified and methods to isolate and identify these elements had been developed. His idea of atoms of different elements being distinguishable by respective relative weights was a new and supportable concept.

> **Ss** | **Science Says**
>
> "The elements of oxygen may combine with a certain portion of nitrous gas or with twice that portion, but with no intermediate quantity."
>
> —John Dalton (1766–1844)

How Atoms Look Today

Obviously, atoms in ancient times never looked any different than they do in the present, but the model began to change when Dalton proposed his atomic theory. Further developments in the modern view of atomic structure occurred in 1897 at

Cambridge University in England when J. J. Thompson discovered the electron with its negative electrical charge. In 1913, Ernest Rutherford discovered that each atom has a positively charged core, or nucleus, which comprises 99 percent of the mass of the atom. The nucleus is itself made up of positively charged particles called protons and neutral particles (not electrically charged) called neutrons. A cloud of negatively charged electrons swirl around the nucleus. Each electron is around 2,000 times lighter than the lightest element, hydrogen.

 Fast Facts

According to the modern atomic theory, bonding between atoms is electrical, and separation between atoms is maintained by the balance of the forces of attraction and repulsion, similar to those demonstrated by the poles of magnets. When electrons are shared between atoms, a "covalent bond" is formed—"co" meaning "together" and "valent" referring to "valence," which is derived from the Latin *valentia*, meaning "strength or capacity."

The modern atomic model was further developed by Danish scientist Niels Bohr, who received the Nobel Prize in Physics in 1922 for the development of quantum mechanics. Bohr and others established the concept of the atomic orbital, which is a confined region of space around the nucleus. According to the Bohr model, there are clearly defined orbitals, representing different levels of energy, that an electron can occupy. For an electron to move from one orbital to another it must either absorb or emit a particle of light known as a photon. Chemical properties of an element are determined by the electrons orbiting the nucleus. Since the electrons are in constant movement, atoms cannot be thought of as "solid" in the sense of a wall in your house, but because the electrons move so fast, they can be modeled as a solid shell. In contrast to the indivisible solid atoms envisioned by Greek philosophers and by John Dalton, "modern" atoms are characterized by constant movement and change!

Fast Facts

Although a few gaseous elements, including helium and argon, normally exist as individual atoms, most of the elements do not. Some, such as hydrogen and oxygen, form molecules of two (or sometimes more) atoms. Most metals exist as arrays of many atoms arranged in a three-dimensional pattern.

Motion—Everything Moves Like Clockwork

While the Greek philosopher Aristotle made many discoveries through careful observation, his description of motion was later proved to be flawed. He proposed that a force is necessary to keep an object in motion and that the speed of a falling object is proportional to its weight. Although these ideas were accepted for almost 2,000 years, they were both disproved by the seventeenth century.

The first crack came with observations of objects in the sky by Polish nobleman Nicolaus Copernicus. His *On the Revolution of the Celestial Spheres*, often credited with beginning the Scientific Revolution, was published in 1543. In it Copernicus proposed that the sun, not Earth, was the center of the solar system. Based on this idea and careful observation of the planets, German astronomer Johannes Kepler developed his laws of planetary motion:

1. The orbit of every planet is an ellipse with the sun at one of the foci (foci being "focal points").

2. A line joining a planet and the sun sweeps out equal areas during equal intervals of time.

3. The squares of the orbital periods of planets are directly proportional to the cubes of the semi-major axis of the orbits.

Our modern understanding of motion came together with the publication in 1687 of Isaac Newton's three laws of motion in his *Philosophiae Naturalis Principia Mathematica*:

1. Every object in a state of uniform motion remains in motion unless an external force is applied to it.

Ff **Fast Facts**

Newton's equation for motion is: $F_{net} = m \times a$. F_{net} represents the total of all forces acting on an object. Mass is represented by m and acceleration is a. Thus, if you multiply the mass times the acceleration, you know the total of forces in play. This is important because to apply any of Newton's laws of motion you need to know the net force.

2. The acceleration of an object is equal to the force applied divided by the mass of the object.

3. When one object exerts a force on a second object, the second object exerts an equal and opposite force.

These three laws form the basis of all of the science and engineering principles of motion. Newton's laws of motion completely supplanted the idea that a force is needed to keep something moving.

Us **Uncommon Sense**

Some people still believe, like Aristotle, that a force is necessary to keep an object in motion. This is directly contradictory to Newton's first law of motion. For example, a meteor falling into Earth's atmosphere does not run out of energy. The presence of the force of friction with Earth's atmosphere causes it to slow. If our planet had no atmosphere, the meteor would not slow at all.

Gravity

It was Galileo Galilei who disproved Aristotle's idea that heavier objects fall faster than lighter objects. Although most historians discount the story that he dropped objects of different masses from the Tower of Pisa, Galileo did determine that falling objects accelerate at the same rate, regardless of their mass. Isaac Newton later proposed an attractive force between Earth and other objects, which also acted between the sun and the planets.

The famous story of Newton under the apple tree and an apple falling on his head is probably not true, but by observing an apple fall he realized it went from zero acceleration to the speed when it hit the ground. What force caused the acceleration? It was not, as the ancient Greeks felt, a solid body seeking its "natural place." Newton realized that because of gravitational attraction, the moon was continually falling toward Earth, but its acceleration in orbit formed a balance. Based on these attractions between bodies, Newton developed the law of universal gravitation.

Ff **Fast Facts**

Gravitational force between two bodies whose masses were known was first measured by Henry Cavendish in 1798. Cavendish also discovered hydrogen, which he called "inflammable air" in his treatise "On Factitious Airs" in 1766. From 1797 to 1798, a Cavendish experiment was the first to measure gravity between masses—two 350-pound lead spheres and two 1.61-pound lead spheres. His aim was to determine the density of Earth. Although he did not get an accurate measurement, his work allowed others to do so later.

If you ever wondered why astronauts in orbit are weightless, it's simple—they're falling at the same speed of their spacecraft. The acceleration of the craft counterbalances the pull of gravity. And without the balance that gravity affords to objects in motion, the universe simply would not hold together. Had Newton not demonstrated that all bodies attract each other gravitationally and that the force depends on the product of two masses divided by the square of the distance separating them, no one would have ever had a space program.

Ff **Fast Facts**

There is a difference between the mass of an object and its weight. Mass is a measure of the amount of material that an object has. Weight is a measure of the gravitational pull on the object, so weight is proportional to mass. However, the weight of an object depends on its relationship to Earth or some other body. The weight of a person standing on the moon is about one-sixth the weight of the same person on Earth. The person's mass, however, does not change.

Germ Theory

If you were to ask a number of medical experts to name the most important theory ever developed in medical science, it is likely that you would get the same answer from all of them: the germ theory of disease. While it may seem obvious today that many diseases are caused by microorganisms, the germ theory is actually a fairly recent concept.

As recently as 200 years ago, doctors had to work with no real understanding of the cause of illness or how it is transmitted from one person to another. Ancient explanations relied on the supernatural. Later explanations relied on spontaneous generation, the idea that living things can rise from nonliving matter. In this view disease was spontaneously generated instead of being created by microorganisms that grow and reproduce.

Ff **Fast Facts**

Although it has been more than 150 years since Ignaz Semmelweis found that hand washing reduces the spread of infection, the message must be repeated frequently. A Johns Hopkins research team reported the results of a study in 2004, recommending that doctors wash their hands frequently and thoroughly in order to control the spread of methicillin resistant staph infections (MRSA).

Microorganisms were first directly observed by Anton van Leeuwenhoek, who is considered the father of microbiology, in the 1670s. The connection to disease did not come about for quite some time after his discovery, though. A major breakthrough occurred in 1847 when Ignaz Semmelweis, a Hungarian obstetrician, observed a high incidence of death from puerperal fever among women who delivered babies in hospitals, while the disease was relatively rare for home births. Recognizing that some sort of contagion was at work and that doctors might be transmitting the disease, he began insisting the doctors wash their hands before examining their patients. Fatalities from the disease dropped from 30 percent of births to 2 percent.

In the 1860s, Louis Pasteur demonstrated that fermentation and the growth of microorganisms in nutrient broths did not proceed by spontaneous generation. By heating the broth, he destroyed microorganisms and found that they did not begin growing again if the broth was sealed from contact with air. This process, known as pasteurization, is still used in the food industry. Further research led Pasteur to the discovery that certain diseases are caused by microorganisms.

About 10 years later, Robert Koch devised a series of experiments to verify the germ theory of disease. Koch demonstrated that anthrax was caused by the bacterium *Bacillus anthracis.*

The germ theory of disease proposes that microorganisms are the cause of many diseases. Although highly controversial when first proposed, it is now a cornerstone of modern medicine and clinical microbiology, leading to such important innovations as antibiotics and hygienic practices.

Even today, some diseases previously thought to be of genetic or environmental causes have been discovered to be caused by microorganisms. The human papilloma virus (HPV) is now known to be capable of causing cervical cancer. Hepatitis B or C has been shown to be a cause of liver cancer. The discovery that stomach ulcers are caused by a bacterium revolutionized the treatment of this disease.

Ff **Fast Facts**

The Gardasil vaccine, first approved for American use in 2006, was tested on 11,000 females aged 9 to 26 around the world before its release. Contrary to emotional opinions that arose when Governor Rick Perry of Texas made the human papilloma virus–fighting vaccination mandatory for girl students, there were no serious side effects noted during testing. In October of 2007, Great Britain made the HPV vaccine free for all women over age 12.

Evolutionary biologist Paul Ewald hypothesizes that many diseases not currently considered infectious are caused by microorganisms. For example, heart disease can be linked to *Chlamydia pneumoniae*, a bacterium known to cause pneumonia and bronchitis. He thinks that someday, many diseases currently thought to have a genetic cause may be curable by isolating and eradicating bacterial agents at the root of the problem.

Ss **Science Says**

"Surprisingly, neglect of the germ's-eye view of the world is not restricted to the average person; it extends to medicine as a whole for most of its history. Only during the past twenty years have researchers emphasized the importance of looking at a germ's evolutionary scorecard. This scrutiny is suggesting solutions to the most damaging problems of medicine as well as the most irritating. Both categories of problems are important."
—Paul Ewald

Plate Tectonics

In 1912, Alfred Wegener, a German meteorologist, proposed that all the continents of Earth had once been joined in a supercontinent he named Pangaea. He presented various evidence including the ancient fern Glossopteris, which could be found in Africa, Australia, India, and South America. He showed rock strata similarities in these continents, as well as glaciation markings, and posited that shifts in the southern magnetic pole could only be accounted for by the gradual splitting of this giant land mass.

In 1962 it was proposed that Earth's continents were composed of lithospheric (solid outermost layer) plates that move slowly across the asthenosphere, the hot rock zone in the upper mantle below Earth's crust. Scientists determined that at midocean ridges, long cracks from which lava rises, the two sides move apart and thus plates diverge and have new material added to them.

In 1965, Canadian geophysicist J. Tuzo Wilson introduced the term "plate" to describe the broken segments of the former supercontinent. Two years later, American geologist W. Jason Morgan suggested that there are 12 plates making up Earth's surface, and shortly thereafter French geophysicist Xavier Le Pichon published a paper revealing the location and type of plates and the direction in which they are moving.

The Big Bang

According to the most widely accepted theory of the beginning of the universe, the Big Bang theory, the universe began when all the matter of the universe exploded instantaneously from a single point. The "big bang" expansion from this point set in motion the process in which the universe expands and matter cools along the way. The big bang should not be conceived, however, as a phenomenon akin to a bomb going off. Instead it is an expansion that has been going on for billions of years and continues today. For the first million years or so of the universe's existence, it was too hot for atoms to form. As the universe expanded, matter condensed into atoms which were pulled together by gravity to form the first stars and galaxies. Along with the expansion, the universe as a whole became cooler.

A fairly recent piece of evidence for the big bang is cosmic microwave background (CMB) radiation. CMB is a form of electromagnetic radiation that fills the whole universe. Characterization of the CMB indicates that the average temperature of the universe is about 3 degrees above absolute zero. This information has helped scientists determine that the universe is about 14 billion years old. Interestingly, the term "big bang" was first used in 1950 by astronomer Fred Hoyle, a critic of the theory. While he intended it to be a derisive term, it has since been adopted as the consensus name of the theory.

Chapter 22

Some Important Scientists

"Discovery consists of seeing what everyone else has seen and thinking what no one else has thought."

—Albert Szent-Geörgyi (1893–1986)

It would be impossible to list all the important people who have made major contributions to science in this chapter. While some names are familiar, almost household words—Newton, Edison, Curie, Einstein— you may never have heard of other scientists who have made equally important contributions. Can you name any famous female scientist other than Madame Curie? Read this and you'll discover a woman whose contribution to computing helped make your personal computer possible. Do you know how the scientific method came about? It happened in the Middle East, but probably not in the place you'd imagine. And how about zero, the concept that expanded mathematics? It started with a dot in that place called India.

Western Science Began in Greece

Although the modern concept of science is only about 500 years old, ancient Greek philosophers laid the foundation of Western scientific ideas. Aristotle is perhaps the most famous and influential of these philosophers, but there were many others, including Democritus (Chapter 21), who developed the idea of a smallest particle of matter. Many science historians attribute the beginnings of modern science to Thales of Miletus, who presented rational explanations for physical phenomena that were nonreligious in nature, arguing, for example, that lightning bolts were not tossed from Mount Olympus by the god Zeus. Here is a short list of Greek scientists worth knowing about.

◆ Archimedes stated: *The apparent loss in weight of a body immersed in a fluid is equal to the weight of the displaced fluid.* Although his famous statement about being able to move the world with the proper fulcrum was not practical, he invented a machine that allowed shipbuilders to lift ships from a dock down to the sea.

◆ Eratosthenes calculated the circumference of Earth by comparing the angle of a pole's shadow at Alexandria, Egypt, at noon and the sun being directly overhead a well in Syene in the southern part of the country during the summer solstice.

Ff Fast Facts

Eratosthenes recognized that the sun was directly over Syene, Egypt, on the summer solstice and that Alexandria was due north of Syene (which is Aswan in modern Egypt). He knew from a previous measurement that in Alexandria at that exact moment, the angle of elevation of the sun would be $\frac{1}{50}$ of a full circle (7°12') south of the zenith. Using the distance between the cities (5,000 "stadia" by Greek standards), he multiplied 5,000 stadia by 50 and determined Earth's circumference to be 250,000 stadia, which isn't far off the modern circumference measurement of 24,900 miles.

◆ Euclid made mathematic knowledge systematic and developed mathematical principles that are still important today, the most prominent being his proof that the amount of *prime numbers* is infinite.

De Definition

A **prime number** is any number greater than 1 that can be divided only by the number 1 and itself, such as 2, 3, 5, 7, and 11. Prime numbers are important; many algorithms to provide computer cryptography are based on very large prime numbers.

♦ Galen went from treating gladiators to being the court physician of Roman emperors like Marcus Aurelius. Prior to Galen, physicians thought the arteries in the body carried air, not blood. He also mapped out the majority of cranial nerves. His writings were so influential they basically went unquestioned for a thousand years.

♦ Hippocrates of Cos gave us the Hippocratic Oath and proved that diseases had a logical and rational cause that could be determined by observing everything about a patient.

♦ Ptolemy's astronomical book, the *Megalê Syntaxis* ("Big Explanation"), summarized astronomical knowledge in the second century, and remained the most important Western work on the subject for 14 centuries following.

The Indians Were There First

Much of science depends on math, and according to scholars in the country of India, the math the world uses today originated there long ago. Indian school texts attribute the invention of zero to a fellow name Aryabhatt, born in 476 C.E. His text on mathematics entitled *Aryabhatiyam* dealt with calculating the motion of the planets and the time of eclipses. According to Indian scholars, Aryabhatt put forth the concept of zero and calculated the value of pi to its commonly expressed value—3.1416.

Another important ancient Indian scientist was Bhaskaracharya, whose work *Siddhant Shiromani* describes mathematical techniques and includes many discussions on astronomy. He is considered a genius in algebra, arithmetic, and geometry, as well as being knowledgeable about Earth, including determining that at the poles of the planet there are roughly six months of day and night. His determination of the circumference of Earth was astonishingly close to modern calculations.

> **Ss** | **Science Says**
>
> "Objects fall on earth due to a force of attraction by the Earth. Therefore, the Earth, planets, constellations, moon, and sun are held in orbit due to this attraction."
>
> —Bhaskaracharya (1114–1185)

Historic record indicates that the decimal system came to the West via Arab scholars in the ninth century from the translation of the *Brahmasphutasiddhanta* by the Indian scientist Brahmagupta. Prior to adopting this numeric system from the Arabs, Europeans were using Roman numerals.

Masters of the Middle Kingdom

Although the West is still becoming accustomed to Chinese medicine techniques such as acupuncture, ideas and practices that originated in ancient China have been transforming the world for millennia. Since so many scholars and scientists worked under the rule of emperors, individual scientific achievement by the Chinese can be hard to determine. For example, the Great Wall of China was a magnificent structural achievement, constructed between 220 and 200 B.C.E. under the first Chinese Emperor Qin Shi Huang, but its architect remains unknown.

We do have records, however, showing that the first seismograph for measuring earthquakes was invented by Zhang Heng (78–139 C.E.). It was an instrument in which balls would drop into an urnlike container and reveal the time and direction of an earthquake. And supposedly, a man named Zai-Lun (50–118 C.E.) first invented paper for writing; he was recognized on a 1962 Chinese stamp.

During the Tang Dynasty 618 C.E. to 907 C.E., considered the golden age of early China, four of the most important inventions of all time came into regular use. The compass, gunpowder, papermaking, and printing all had earlier origins but were not commonplace. These items changed China and then the rest of the world as knowledge traveled from China through the Arab world and then to Europe and beyond.

Us | Uncommon Sense

While Johannes Gutenberg is credited with the invention of movable type in 1455, Bi Sheng in China invented movable type in the eleventh century. The European woodblocks used for printing in the fourteenth century used a technique similar to Chinese woodblocks.

Other advances attributed to scholars and inventors of the Tang Dynasty include cast iron, dry docks, the horse collar, matches, the iron plough, and even the parachute. The Tang Dynasty was largely a golden age of literature and art, an inspiring, progressive, and stable society. The creation of woodblock printing made information (with illustrations) available to a great many more people.

During this period, in 725 C.E., the Buddhist monk Yi Xing invented the world's first clockwork escapement mechanism, which had a bell struck automatically every hour; this was the precursor to modern clocks.

In the centuries following, during the Song Dynasty (960–1279 C.E.), many other technological breakthroughs were achieved in China. One important scientist during this period was Shen Kuo, who discovered the concept of true north and calculated the position of the pole star that had shifted over the centuries. Another notable creation was the "pound lock" used in canals and rivers; this invention made the Panama Canal possible. Although the mechanism may have been used by Romans and others, the version invented in 984 C.E. by engineer Qiao Weiyo in Huainan was a prototype for the type of locks in current use.

Ff **Fast Facts**

The pull of gravity from the sun and moon acts on the equatorial bulge of Earth that arises during its rotation. Because of this, Earth's axis wobbles in its orbit, covering a circle in a period of 26,000 years. This phenomenon, known as precession, affects the equinoxes and solstices, because the background stars in the sky do not remain constant. Our current pole star, Polaris, which can be seen by observing the alignment of two stars in the "Big Dipper," was just another star in the sky in ancient times. Four thousand years ago, the pole star was Thuban in the Draco the Dragon constellation.

Arabic Numbers and the Scientific Method

Arab scholars advanced many concepts in mathematics, including the concept of zero and the development of algebra. The idea of zero is so ingrained in our system of using numbers that it is impossible to picture it as anything other than a natural concept, apparent to everyone. It seems strange that neither the Greeks nor the Romans used the concept, nor did European cultures after the demise of the Roman Empire. It was an Italian mathematician, Leonardo of Pisa (more commonly known as "Fibonacci"), who introduced Arabic numerals to Europe in the Middle Ages. The Persian mathematician Al-Khwarizmi, considered to be the father of algebra, wrote the first book on the systematic solution of linear and quadratic equations. Although Al-Khwarizmi is often credited with the invention of the concept of zero, he mentions his source in his book *On the Calculation with Hindu Numerals*, which was written about 825 C.E.

Us **Uncommon Sense**

Most people think that multiplication is impossible with Roman numerals, particularly since they had no zero. Of course, they did multiply. The techniques for multiplication with Roman numerals is explained in detail on some websites. Have a look at www.legionxxiv.org/numerals and www.phy6.org/outreach/edu/roman.htm to get the idea. In short, the Romans' method boiled down to binary notations, with each digit representing a power of 2.

A polymath (someone distinguished in several fields of study) named Ibn al-Haytham (also known as Alhacen, 965–1039 C.E.) pioneered both modern optics and the scientific method. Muslim scientists kept careful records, and the experiments noted by Ibn al-Haytham in his *Book of Optics* published in 1021 is recognized as the beginning of the modern scientific method. The evidence established by his experiments were systematically repeated to prove that ancient Greek ideas that (a) human eyes emit rays of light (an idea supported by Euclid), and (b) that physical objects emit physical particles which are received by the eyes, were both wrong. The steps he used are the same scientists employ today:

1. Observation

2. Statement of the problem

3. Formulation of the hypothesis

4. Testing of the hypothesis using experimentation

5. Analysis of experimental results

6. Interpretation of data and formulation of the conclusion

7. Publication of findings

Hard as it may seem to believe that people thought otherwise, Ibn al-Haytham was the first to prove that light travels in straight lines, which he did over many years, using methods like placing straight sticks or threads pulled tight next to beams of light streaming through pinholes into dark rooms. Vision, he demonstrated, was about the eye perceiving these rays of light traveling in straight lines. Ibn al-Haytham's work in optics helped later in the development of the telescope, so in many ways, this scientist's life work was far-reaching.

Fast Facts

In psychological terms, visual perception is the ability to interpret the information from visible light that reaches the eyes. By explaining in Book III of his *Book of Optics* that vision truly occurs in the brain, rather than the eyes, Ibn al-Haytham became the first scientist to venture into the psychology of sight. Understanding of what is seen, he demonstrated, depends entirely on the experience of the person doing the observing.

A Short List of More Recent Scientists

So far, this chapter has discussed names which, with the exception of some of the Greeks, are probably unfamiliar to most readers. A fair amount of discussion was necessary to establish the context of times and cultures. The rest of this chapter will mention people who may be more familiar. There is no intention to give them short shrift, but each only receives too-short acknowledgment. The following list includes a few notable scientists of the past few hundred years. Others have been mentioned in previous chapters. It is not comprehensive, though.

- ◆ Leo Baekeland (1863–1944)—Chemist who invented the first commercial plastic, Bakelite. Bakelite was used extensively for radio, telephones, and electrical insulators.

- ◆ Alexander Graham Bell (1847–1922)—The inventor of the telephone, who also invented a metal detector and had notable contributions in aeronautics.

- ◆ Marie Curie (1867–1934)—Polish physicist and pioneer in the field of radioactivity. She was twice honored as a Nobel laureate and one of only two people to win in two different sciences.

- ◆ Charles Darwin (1809–1882)—English naturalist whose 1859 book *The Origin of Species* established evolution as the dominant scientific explanation of species in nature.

- ◆ Thomas Edison (1847–1931)—American inventor who is most famous for the phonograph and light bulb. One of the first to apply principles of mass production to the process of invention, he received a record 1,093 U.S. patents.

◆ Albert Einstein (1879–1955)—German-born theoretical physicist best known for the theory of relativity and the mass-energy equivalence formula, $E = mc^2$. Einstein's special theory of relativity reconciled mechanics with electromagnetism. His general theory of relativity, by extending relativity to nonuniform motion, created a new theory of gravitation and predicted the later proven gravitational bending of light.

◆ Michael Faraday (1791–1867)—English physicist and chemist who discovered diamagnetism, electromagnetic induction, electromagnetic rotation, field theory, and the magneto-optical effect.

◆ Enrico Fermi (1901–1954)—Italian physicist noted for his work in the development of the first nuclear reactor and developments in quantum theory.

◆ Sir Alexander Fleming (1881–1955)—Scottish biologist and pharmacologist whose many achievements include the discovery of the antibiotic penicillin in 1928.

◆ Rosalind Elsie Franklin (1920–1958)—A pioneer molecular biologist responsible for much of the work that led to the understanding of deoxyribonucleic acid (DNA).

◆ Sigmund Freud (1856–1939)—Austrian neurologist and psychiatrist who was the founder of psychoanalysis.

◆ Galileo Galilei (1564–1642)—An Italian scientist who formulated the basic law of falling bodies and is viewed as the father of modern physics.

◆ Rear Admiral Grace Murray Hopper (1906–1992)—American computer scientist and United States Navy officer who developed the first compiler for the computer programming language COBOL and popularized the idea that programs could be written in language close to English rather than in machine code used at the time.

◆ Edwin Hubble (1889–1953)—American astronomer who discovered the red shift of galaxies and developed the idea of an expanding universe.

◆ Joseph Lister (1827–1912)—An English surgeon who first promoted the idea of sterile surgery. (Listerine mouthwash is named for him.)

◆ Guglielmo Marconi (1847–1937)—An Italian inventor known for inventing the radio (although Nikola Tesla may have done it first).

- Samuel F. B. Morse (1791–1872)—American creator of the telegraph and co-inventor, with Alfred Vail, of the Morse code.

- Isaac Newton (1643–1727)—English physicist and mathematician most noted for his laws of motion and of gravitation. Newton also shares credit with Gottfried Leibniz for the invention of the calculus.

- Alfred Nobel (1833–1896)—Swedish chemist and businessman who invented dynamite and established the Nobel Prize.

- J. Robert Oppenheimer (1904–1967)—American theoretical physicist directed the Manhattan Project and was known as the "father of the atomic bomb."

- Nikolaus Otto (1832–1891)—German inventor of the first four-stroke internal-combustion engine that was the basis of current automobile motors.

- Louis Pasteur (1822–1895)—French chemist and microbiologist who confirmed the germ theory of disease, developed a process for making milk safe, and created the first vaccine for rabies, among his many achievements.

- Linus Pauling (1901–1994)—American scientist who was one of the first to work in the fields of molecular biology, orthomolecular medicine, and quantum chemistry, and who is the only person other than Marie Curie to receive a Nobel Prize in two different fields.

- Vera Rubin (1928–)—American astronomer who has done work on the rotation rates and structures of galaxies. Her discoveries have provided evidence of dark matter.

- Nikola Tesla (1856–1943)—Serbian/American physicist and inventor who made substantial contributions in electricity and magnetism, including alternating electric current and AC motors.

- Werner von Braun (1912–1977)—German scientist who developed rocket technology in both Germany and the United States and is seen as the father of the U.S. space program.

- James Watson and Francis Crick (1928–   ; 1916–2004)—Discoverers of the DNA molecular structure, probably the most important biological discovery of the twentieth century.

- James Watt (1736–1819)—Scottish inventor and engineer who improved the steam engine and accelerated the Industrial Revolution.

♦ Orville Wright and Wilbur Wright (1871–1948; 1867–1912)—American brothers who invented and built the world's first successful airplane and made the first sustained heavier-than-air human flight on December 17, 1903, at Kitty Hawk, North Carolina.

It is quite likely that the previous list missed at least one scientist you would have mentioned. Entire books and even encyclopedias have been written without mentioning everyone who has contributed to science. Scientific research goes on every day around the world. Millions of scientists contribute to our lives, working in universities, manufacturing companies, hospitals, and many other places.

"Let us be clear about it. What science can do, it will do, some time, somewhere, whatever obstacles may be put in its way."

—Christian de Duve (1917–)

Chapter 23

Some Unanswered Questions

"For every one billion particles of antimatter there were one billion and one particles of matter. And when the mutual annihilation was complete, one billionth remained—and that's our present universe."

—*Albert Einstein (1879–1955)*

An interesting aspect of science is that you never end up with all the answers. Every discovery and every advance brings with it new questions, so there is always another problem to attack.

Some of the unanswered questions are universal in scale. For example: What is the universe made of? Others are smaller in scale but more important in our daily lives: How do we think?

What is the universe made of?

This is one of the questions that seems like it should have an obvious answer, but it doesn't. We know that there are about 100 elements and that any other elements that might exist are very unstable. Elements themselves are made up of electrons, neutrons, and protons, which together form atoms. And while neutrons and protons form the nucleus around which electrons spin, both neutrons and protons are composed of subatomic particles called quarks. It takes three quarks to make up either a neutron or a proton.

When we look at stars and galaxies we can detect many of the elements, but almost everything that we see is hydrogen or helium. So the universe is made of the elements that we know about, primarily hydrogen and helium, right? Wrong.

The problem is that there has to be more to the universe than what we can see. In the 1960s, astronomers discovered that galaxies would fly apart unless they had more mass than what we can measure in their stars. They proposed the existence of dark matter, which we cannot see but which exerts a pull of gravity. There is also not enough known matter to account for the distribution of galaxies in the universe. Cosmologists have concluded that the gravitational forces that shape the universe must come from another, so far undiscovered type of matter. They estimate that this exotic dark matter accounts for about one-fourth of the universe—and we really don't know anything about it other than its effects on the matter we see.

The most recent data we have about the composition of the universe comes to us via the Wilkinson Microwave Anisotropy Probe (WMAP), an Explorer mission from the National Aeronautics and Space Administration. In 2008, after studying cosmic microwave background (CMB) radiation—the "afterglow" light left over from the big bang—the WMAP revealed that the oldest light has been traveling across the universe for about 13.7 billion years. Data from the WMAP project also demonstrated that the entire mass density of the universe is equivalent to 5.9 protons per cubic meter. However, the density of the atom itself is roughly one proton per 4 cubic meters. Calculations show that atoms make up only 4.6 percent of the density of the universe.

Now let's add another complication. We know from observations of distant galaxies that the universe is expanding. We also know that the force of gravity between any two chunks of matter in the universe should slow the rate of that expansion as galaxies pull on one another. For decades, scientists have been working to determine the rate at which the expansion of the universe is decelerating. That would give us a good handle on the amount of matter that is trying to pull itself together. WMAP has provided data to help answer that question. There are never easy answers, though. We find that the rate of expansion is actually accelerating. That means there is something working against gravity to push the universe apart—not at all what was expected. Cosmologists now theorize that there is a form of energy, which they call dark energy, that is pushing on the universe. No one knows what it is but only that its effects appear when we measure the universe. Ordinary matter and dark matter combined account for only about 30 percent of everything. The rest is dark energy (remember, $E = mc^2$—matter and energy are two faces of the same thing).

So, to answer the question "What is the universe made of?" we will first have to answer other questions:

1. What is dark matter (and where is it)?

2. Is there more than one kind of dark matter?

3. What is dark energy?

Does anything exist outside the universe?

The short answer to this question is, "Nobody knows." So let's start with the anthropic principle, a physics term that states human existence has built-in observational constraints based on our human condition. In short, the only universe humans will see is one that supports life, because if it did not support life, we could not exist to observe that universe. The term was coined in 1973 when theoretical astrophysicist Brandon Carter proposed it to a Kraków, Poland, symposium honoring the 500th birthday of Copernicus.

This is the kind of question that is hard to test. The first hurdle is to define the "universe." If you go with the original meaning of universe, that is, everything that exists, then by definition nothing exists outside the universe. That of course is much too simple a way to answer the question.

What if you define the universe as everything that we can possibly observe? Now you have a more interesting question. If it is outside the universe, we can't observe it, so theories have to be based on indirect evidence and even some philosophical arguments. We know that the universe is expanding, so it is natural to wonder what it is expanding into.

One way of looking at that is that the universe defines time and space, so it is not expanding into anything because time and space do not exist outside the universe.

Another theory proposes the idea of a multiverse—multiple universes existing with varying characteristics, shifting and reforming like the baubles inside a kaleidoscope. Expanding on that is the idea of fecund universes that "give birth" to other universes.

On the whole this question may be more philosophical than scientific because it is not based on evidence. Some researchers, though, are looking at the big bang and trying to find evidence to show that something existed before it occurred and that we may be able to learn about things outside our universe based on the interactions of its material before the big bang. In the meantime, this may be the hardest question of all to answer.

Is there anyone else out there?

The question of whether intelligent life exists elsewhere in the universe has been around for a long time. Hindu scriptures speak of innumerable universes created by the Supreme Personality of Godhead, while the Jewish Talmud states that there are at least 18,000 other worlds. To examine the question in nonreligious terms, we should keep in mind that the human body is made up of 95 percent hydrogen, oxygen, and carbon atoms. So to find life *like our own*, we would probably need to find planets where similar molecular configurations could exist. However, the possibility of silicon-based life forms has been proposed, and some scientists think ammonia could do for other life forms what water does for life on Earth.

Ss | Science Says

"What a splendid perspective contact with a different civilization might provide! In a cosmic setting vast and old beyond ordinary human understanding we are a little lonely, and we ponder the ultimate significance, if any, of our tiny but exquisite blue planet, the Earth … In the deepest sense the search for extraterrestrial intelligence is a search for ourselves."

—Carl Sagan (1934–1996)

Most scientists involved in biology beyond Earth think that complex multicellular life, as found on Earth, would be a highly improbable circumstance on most planets. On the other hand, in a universe of hundreds of billions of galaxies, each made of hundreds of billions of stars, the highly improbable could happen many times. However, calculating the possibilities and finding evidence are two very different things.

The Search for Extra-Terrestrial Intelligence (SETI) program has existed for four decades and so far it has not turned up a radio signal from another world. Nevertheless, interest is not waning. The European Space Agency launched the Darwin mission designed to find Earth-like planets, and the French Space Agency's COROT mission was launched in 2006 with a similar mandate. At the time of this writing the Kepler mission is scheduled for launch by NASA in November 2008.

To date, however, scientists have only identified a few dozen exosolar planetary bodies. Exosolar means planets circling stars, not necessarily planets that might support living beings. So for now, the answer to the question of "other life out there" remains "there is no evidence for it."

Ff **Fast Facts**

On April 24, 2007, Chilean scientists at the European Southern Observatory announced they had discovered Gliese 581 c, the first Earth-like planet orbiting within the habitable zone of its star (in this case, Gliese 581, a red dwarf). It was initially thought that this planet could contain liquid water. Subsequent computer simulations by a team in Germany's Institute for Climate Impact Research have suggested that gases in that planet's atmosphere raised the surface temperature above the boiling point of water. It was a nice idea while it lasted.

How do we think?

Using functional magnetic resonance imaging (fMRI), scientists can see what happens inside the brain while thought is taking place. fMRI is a neurological imaging technique that uses the level of oxygen in the blood to show what structures of the brain are active during given mental operations. By studying activity in the visual and intraparietal cortexes (the sensory processors in the brain), they have discovered that the way people think may be shaped by the evolutionary history of the species and the way our brains developed. Even our moral outlook, some scientists now feel, might have something to do with how we are "wired."

Our brains are composed of nerve cells known as neurons that connect through signal-receiving dendrites and signal-transmitting axons. Imagine a leafless tree during winter and you have an idea of these branchlike projections. When dendrites and axons connect they do not touch. Rather, a gap called a synapse fires and an electrical impulse is sent. Each neuron has thousands of synapses, and each transmission takes place at lightning speed (about 300 milliseconds), so with the billions of neurons in the brain operating at such a clip, there is an infinite capacity for storing information, and endless neuron pathways.

Ff **Fast Facts**

Apparently, we can only think so fast. In 2004, three theoretical neurophysicists from the Max Planck Institute for Flow Research in Göttingen, Germany, discovered an upper limit on the speed of thought. The researchers used a mathematical model to show that due to the mechanical limitations of neural connections, the switching has an upper limit. The only way to beat this "speed limit" would be if every neuron in the brain could manage to be connected to every other single neuron.

Does this capacity diminish with age? The good news is that the brain can change. Thomas Elbert, a professor of psychology at the University of Konstanz in Germany, who has published extensively on brain activity, has demonstrated that the adult brain has malleable plasticity rivaling that of a child. This means that if we keep our brains healthy and active they will remain flexible, and our capacity for learning can last as long as we are living.

Ff **Fast Facts**

Neurogenesis is the process by which neurons are born, and now it may be possible to grow new cells in the adult brain. In 2000, a Stanford study revealed in the magazine *Science* that transplanted bone marrow cells can migrate to the brain and turn into neurons. Better yet, in November 2006, Fulton Crews, a professor of pharmacology and psychiatry and director of the Bowles Center for Alcohol Studies at the University of North Carolina at Chapel Hill, spoke at the National Institute on Alcohol Abuse and Alcoholism and described how heavy physical exercise increased neurogenesis in rats. Since humans have a similar brain structure, it follows that new brain cells can literally be exercised into existence.

However, understanding the processes of neurons does not answer another important question: What is the mind and how does it relate to the brain? The description of brain processes sounds remarkably like that of computer processing. There is a big difference between a computer and a human, though. No one has ever built a computer that is self-aware and makes decisions by its own will. Why are people able to think rather than just process information? That question remains unanswered.

How can we make replacement cells for body parts?

Can we make body parts to replace those that wear out during the course of living? This is a popular technique in science fiction to prolong life. But how can it be done—or can it be done at all?

Some researchers think that blueprints for new body parts can be printed out. A process called bio-printing was developed in 2006 by a research team lead by Julie Jadlowiec-Phillippi, a bioengineer at Carnegie Mellon University and Children's Hospital of Pittsburgh. In a meeting of the American Society for Cell Biology in San Diego, she explained that her team had, in conjunction with the university's Robotics Institute, created an inkjet printer to spray a chemical mix onto protein-coated slides, allowing them to manufacture bone and muscle cells in a petri dish. Within the dish were stem cells taken from the muscles of adult mice. While the team did not believe a method for growing body parts was imminent, their "bio-ink" method offered great promise and another method for studying how stem cells mutate into other specialized cell types.

Two years later, a team led by Dr. Anthony Atala at Wake Forest University was manufacturing body parts. When interviewed by CBS News in February of 2008 in a segment called "Growing Miracles," Atala revealed that they had made 18 different types of tissue. Viewers were shown a functioning, engineered heart valve that was to be transplanted into a sheep. Their method sounded simple; they isolated cells capable of regeneration and coaxed them to grow. Like the Carnegie-Mellon team, the Wake Forest team's heart cell regeneration began with the use of an inkjet printer. As the camera rolled, a mouse heart was revealed that was being grown, the cells sprayed on layer by layer.

Ss **Science Says**

"Every cell in your body is programmed to do a job, and our job is to put these cells in the right environment in the lab so they know what to do. To us, it doesn't matter where the cell comes from—whether it's a bladder cell or a blood cell or an adult stem cell—we use whatever cell gets the job done."

—Anthony Atala (1958–)

It wasn't only for animals, however. In an experimental procedure at Thomas Jefferson Hospital in Philadelphia, a patient received a bladder transplant of a bladder grown from her own isolated bladder cells!

At this time, it appears that there may be an answer to this question, although the idea of racks of spare parts is still in the science fiction realm. Will it be possible to someday produce whole organs? That remains a "maybe."

How did life begin?

Inside all living organisms, proteins and nucleic acids interact, causing structures to grow and reproduce. Life produces life in a constant chain of growth. But how did it all begin? When did the first living thing appear on Earth and how did it form since it was not the result of reproduction?

Us **Uncommon Sense**

Charles Darwin's *The Origin of the Species* was the landmark book about evolution, but he did not arrive at his ideas all on his own. French biologist Jean-Baptiste Lamarck (1744–1829) had evolutionary theories that were acknowledged in Darwin's book. Lamarck believed that biological lessons learned by organisms during their lives were the reason species adapted, so that necessary changes would be passed on to off-spring. While many people still believe that learned traits can be passed from one generation to the next, genetic study has shown otherwise.

While this is another unanswered question, there are some hints. In 1828, German chemist Friedrich Wohler synthesized urea, an organic molecule that most thought could only be made by living organisms. This demonstrated that there is no difference between physically produced and organically produced molecules. Prior to this accomplishment, it was generally believed that organic (living) compounds were fundamentally different material from inorganic compounds.

In 1953, Stanley L. Miller, one of the pioneers of exobiology (which preceded astrobiology, the study of life in the universe) and Harold C. Urey conducted an experiment on the origin of life at the University of Chicago. The experiment simulated hypothetical conditions Miller and Urey assumed were present on the early Earth and tested the hypothesis that such conditions favored chemical reactions that would synthesize organic compounds from inorganic elements.

Ammonia, hydrogen, methane, and water were sealed in a sterile array of glass tubes and flasks connected to each other. The water was heated to evaporate it, with sparks fired between electrodes to simulate lightning through an atmosphere. After one week, 2 percent of the carbon had formed amino acids, with over half the amount necessary to make proteins in living cells. DNA and RNA were not formed, but the goo that remained was similar to organic material that can be found on a meteorite—meaning this could be a universal process.

Ff **Fast Facts**

Deoxyribonucleic acid (DNA) contains all the genetic instructions for the development and functioning of living organisms. DNA molecules are akin to a set of blueprints used to construct other cell components. And what blueprints they are! The information stored in a human DNA molecule would fill a million-page encyclopedia.

In 1961, Juan Oro, a professor of biochemical and biophysical sciences at the University of Houston, discovered that amino acids could be synthesized via hydrogen cyanide and ammonia in an aqueous solution. His experiment produced a great deal of adenine, which was highly significant because adenine is an organic compound that is one of the four bases in RNA and DNA. Later experiments revealed that the other RNA and DNA bases could arise through simulated prebiotic chemistry with a reducing atmosphere (a la the Miller-Urey experiment).

These experiments have provided a foundation for a theory of the origins of life. However, the production of the molecules on which life is based is a very different thing from producing life itself. Beyond that, even if life were to be produced in a laboratory, more evidence would be needed to show that it had actually happened that way in nature. As of today, questions about how life began remain unanswered.

"The brain is the last and grandest biological frontier, the most complex thing we have yet discovered in our universe. It contains hundreds of billions of cells interlinked through trillions of connections. The brain boggles the mind."

—James D. Watson (1928–)

Chapter 24

Technology into the Future

"I want to build a billion tiny factories, models of each other, which are manufacturing simultaneously. ... The principles of physics, as far as I can see, do not speak against the possibility of maneuvering things atom by atom. It is not an attempt to violate any laws; it is something, in principle, that can be done; but in practice, it has not been done because we are too big."

—*Richard Feynman (1918–1988)*

Engineering works with science to bring about new technologies. Sometimes it is hard to see the small changes and advances as they happen, but changes are always in the works. To really get a feel for the pace of change over the past few decades, think back fifty years or so. In 1960, the space program was just underway, but the biggest computers were slower than today's pocket calculator. There was a phone booth on every corner because cell phones did not exist. And jet airplanes for commercial travel? There were a few, but most planes used propellers.

Technological changes have transformed our world. As science advances, so does technology. Think about the changes that have occurred since 1960. The amazing thing is that the rate of change is constantly accelerating.

New discoveries and advances provide a framework for future technological growth in many directions. The following sections highlight a few of the fields of technology that will likely be in the news for some time.

Nanotechnology

Physicist Richard Feynman's address to the annual meeting of the American Physical Society at the California Institute of Technology in 1959 marked the beginning of what became the science of *nanotechnology.* His talk, entitled "There's Plenty of Room at the Bottom," can be viewed on the internet at www.zyvex.com/nanotech/feynman. html.

De **Definition**

According to the Center for Responsible Nanotechnology (www.crnano.org), **nanotechnology** is the engineering of functional systems at the molecular scale.

In this speech, Feynman described how the entire Encyclopedia Britannica could easily be written on the head of a pin. He called for better electron microscopes and explained that almost any problem could be solved if we could merely look at structure on a molecular level. At the time, computers were so large they filled entire rooms. Feynman spoke of miniaturizing them and the benefits that would result. He mentioned putting "mechanical surgeons" inside blood vessels to repair the body from the inside. Perhaps most exciting, he envisioned building tiny machines, atom by atom. Feynmann offered cash prizes for people who could achieve feats like making the page of a book so small it could only be read by an electron microscope. With this speech, Feynman got the whole field started, and a mad rush to "the bottom" has existed ever since.

How small is the playing field of nanotechnology? Imagine taking a human hair and reducing it down 50,000 to 100,000 times. You'd be in nanometer territory, one billionth of a meter, roughly the length of three to six atoms placed side by side.

Originally, nanotechnology referred to building working machines including motors, robots, and even computers that would be only a few nanometers wide. However, the meaning has changed a bit over time. Today, nanotechnology refers to manufacturing materials by controlling matter at the atomic or molecular scale—anything smaller than 100 nanometers wide, with novel properties.

Ss   **Science Says**

"In living cells, there are tiny machines that put molecules together to make things like potatoes and trees. People are learning to do this, and when we get good at it, we'll have machines that can make things like solar cells, computers, and spaceships. Like the machines that make the wood in trees, these machines will be able to make things with low cost and almost no pollution."

—Eric Drexler (1955–)

At the nanoscale, materials show different properties and behave differently from larger-scale materials. The website of the National Nanotechnology Initiative (www. nano.gov/html/facts/whatIsNano.html) points out that many natural processes and materials operate at the nanoscale. The strength of a spider web, water repellancy of leaves, and ability of a fly to walk on the ceiling are all nanoscale phenomena.

A key benefit of nanotechnology is that it is an improved manufacturing process. Control at the atomic level means that the process can be more efficient and, ultimately, less expensive. Because the field is so new, no one really knows where it is going. Predictions of its impact range from being able to make new and better tools, such as knives that never become dull, to making tiny, self-replicating robots the size of a grain of pollen—or smaller—that can run around and perform tasks independent from human guidance.

One of the possible applications of nanotechnology is in the manufacture of more efficient computers. Miniaturization has been one of the hallmarks of computer design and nanotechnology will allow miniaturization to unheard-of scales. Computers of the future may compare to today's fastest machines the same way that our computers compare to the giant, plodding, tube-driven computers of the 1950s.

Molecular Medicine

Molecular medicine is another field that uses the small scale in practical applications. It studies and treats diseases on the molecular and cellular level by focusing on the biochemical processes at work. One example of the use of molecular medicine is in the study of catecholamines, which regulate immune and inflammatory responses in the body and regulate the "fight or flight" response. Catecholamines are neurotransmitter chemicals in the brain; they include dopamine, epinephrine (adrenaline), and norepinephrine. Biochemist Julius Axelrod shared a Nobel Prize in Physiology or Medicine in 1970 with Bernard Katz and Ulf von Euler for work on the release and reuptake of catecholamines. The most interesting part of their work was the revelation that adrenaline is recycled (the "reuptake").

Ff Fast Facts

Epinephrine used as a drug treats cardiac arrest and is also used as a bronchodilator for asthmatics. It has a suppressive effect on the immune system. This is why allergy patients receiving immunotherapy are often given an epinephrine rinse prior to the allergen extract.

Axelrod's story is a good example of how molecular medicine is in use today. In 1949, his work at the National Heart Institute on the effects of caffeine led him to an interest in the sympathetic nervous system and the study of epinephrine and norepinephrine. His studies of the function of the nervous system at the molecular level led to development of drugs such as Prozac, which blocks the reuptake of the neurotransmitter serotonin.

An application of nanotechnology in molecular medicine is a system used to perform DNA testing at the benchtop level. Using gold nanoparticles (typically 13–20 nanometers in diameter), the process simplifies molecular diagnostic testing and provides results to hospital laboratories that were formerly available only from large off-site labs. These instruments have diagnostic applications in blood screening, cardiovascular disease, neurodegenerative disorders, and oncology.

Robotics

What comes to mind when you hear the word "robot"? Anyone who has seen the *Star Wars* movies will most likely think immediately of C3PO or R2D2. While these robots make great movie characters, real robots are something completely different. Robots operating in manufacturing facilities generally look just like any other machine. Where they differ is in their controls. While other machines are controlled by a human operator, a robot has some degree of control over its motion. The robot has a computer with software that encodes the instructions for performing particular operations under specific conditions. In other words, a robot is a mechanical device controlled by a computer program that allows it to operate without direct human manipulation.

All robots have three basic parts: a computer, mechanical systems, and electronic systems. The computer is programmed to control the other parts of the robot and make decisions based on input. The mechanical systems are the devices that the robot uses to move and to manipulate objects. The electronic systems act as sensors and carry instructions from the computer to the mechanical and electronic components.

Currently, many robots are used to perform tasks that are repetitive and, therefore, fairly easy to program. They are very common on assembly lines, where a simple task is done repeatedly. The robot is able to determine when a part is in the correct position, and in some cases reposition it if necessary, and then perform the operation. Robots are also used to perform tasks that are too dangerous, or even impossible, for humans to perform. For example, robots have been used to disarm bombs, explore the Chernobyl nuclear reactor, and explore the ocean floor.

Ff | Fast Facts

One of the most amazing uses for a robot is the exploration of Mars. In 2003, NASA sent two robots, the Mars Exploration Rovers, to seek evidence about the history of water on Mars. These robots move across the surface using solar power and send photographs back to Earth. They also carry and use a number of instruments to analyze rocks and soil. The Rovers were initially designed to work for two months before wearing out, although the research team planned to use them as long as possible. At the time of this writing, in 2008, both Rovers continue to explore the surface of Mars and send information home to Earth.

The key difference between a robot and any other machine is the ability to make choices. For example, a remotely controlled toy car, which is operated by radio signals from a controller held by a person, is not a robot. A car that senses obstacles and changes direction to avoid them, on the other hand, could be considered a robot.

Despite how they are presented in movies, very few robots look at all like humans. A major reason for this is that walking on two legs is not a particularly efficient way to move from one place to another. Many robots—for example, welding robots on an automobile assembly line—do not need to move at all. For robots that do need to move from one place to another, wheels or tracks generally provide a stable and efficient platform. In situations where the ability to lift a leg over an obstacle is important, robot designers often use more than two legs. For example, Dante II, a robot built to explore inside volcanoes, has eight legs, closer to the design of a spider than that of a human.

One of the current challenges for robot designers actually has more to do with people than with the robots themselves. As robots are built for more and more tasks, they are more likely to be used by people with no training in dealing with them. Many designers believe that effective interaction will require communication methods that seem natural to humans. Toward this end, there is significant research under-way involving speech recognition software that would allow verbal commands and responses. Nonverbal communication is important in human interactions, so roboti-cists are also working to develop gestures and even facial expressions for robots that would communicate with the general public. Research is even underway to give robots a personality.

New Energy Sources

As human civilizations become more complex, their demand for energy increases rapidly. The earliest agricultural cultures relied on human muscles to provide all the energy needed for food production and locomotion. As more complex societies found an increased need for energy, they harnessed the power of draft animals, the wind, and flowing water. In developed nations today, most energy comes from combustion of fossil fuels—coal, oil, and natural gas—along with smaller contributions from nuclear power and hydroelectric power.

Unfortunately, there are some limitations that accompany these energy sources. The combustion of fossil fuels is accompanied by pollution and is the greatest contributor of greenhouse gases in the atmosphere. Nuclear power generates radioactive waste that will remain dangerous for thousands of years. In addition, fossil fuels and nuclear fuels exist in limited amounts. Although we have not yet reached the end of their availability, we will someday. Hydroelectric power does not pollute but it does alter the flow of rivers as large dams are built. This introduces its own environmental problems. In addition, in many countries, very few suitable places remain in which water power can be tapped.

From the beginning of human history, some of the most important technological developments have involved new ways to produce or harness energy. This is certain to continue into the future. The demand for energy grows with population and advancements in other technologies and the limitations of our current sources means that new energy technologies will be needed. This will be a major focus of future research and development.

Several alternative sources of energy already exist. They are very important in some regions of the world. For example, almost half of the electric power used in Denmark comes from wind power and about 40 percent of Brazil's transportation fuel is ethanol. Overall, however, the world still relies predominantly on fossil fuels for its needs. One of the key considerations for alternative fuels is renewability. Unlike coal or oil, alternative fuels use resources that can be replaced.

Biofuels are produced from plant sources so they can be replaced by growing new crops. Ethanol, the most common biofuel today, can be produced from corn, sugar cane, or other plant materials. The main disadvantage of these fuel sources is that the crop production requires agricultural land that could be used for food production. In addition, the energy used in producing these fuels reduces their overall benefit. Research is underway to find ways to produce fuels from garbage, waste agricultural materials, and wood in order to tap resources that do not compete with food production.

Ff **Fast Facts**

There is at least one country that uses almost no fossil fuels to produce electricity. Iceland produces about three-fourths of its electricity at hydroelectric power plants and about one-fourth from geothermal power plants. In addition, most of the heat for buildings is obtained from hot water from beneath the ground. Iceland is particularly suited for using geological sources of energy because it is located above several volcanoes.

Another research focus is on harnessing the natural motion of air and water to provide energy. In one way, this is not a new technology. Windmills and watermills were major power sources hundreds of years ago. Modern techniques are much more efficient, though. Wind turbines, with blades as long as 100 meters (330 feet), are used to produce electricity, which can then be transported through existing power lines.

Hydroelectric power plants already use the energy of falling water to produce electric power. Small-scale projects already exist that use the power of ocean waves and tides to generate electricity. The idea of wave generators for energy production has been around since the 1930s. Since ocean waves cause an elliptical motion, this can be harnessed to power mechanical motion, generating electricity, which is then piped ashore via undersea cable. There are many plans worldwide to use various designs for power, with the first U. S. plant being built by Pacific Gas and Electric Company off the coast of northern California. It's not a big operation, consisting of only eight buoys, $2\frac{1}{2}$ miles offshore, but when the plant begins operating in 2012, estimates are that it can generate power for as many as 1,500 homes. Similar plans are in place for plants off the coasts of England, Portugal, and Scotland. Although the infrastructure can be expensive, once built, wind and water systems produce electric power using motion that has no cost.

De Definition

Geothermal power is energy that is generated by using the heat stored beneath the surface of Earth. Currently geothermal plants provide less than 1 percent of the energy needs for the world.

Geothermal energy comes from beneath the earth. The planet's interior is heated as radioactive materials break down (the same energy source we use in nuclear power plants). This heat slowly works its way toward the surface. Geothermal power plants use this heat to produce steam to run power plant turbines. Geothermal energy can also be used to heat buildings by pumping warm water into a heat pump and transferring its energy into the surrounding air.

One of the oldest sources of energy is once again becoming a prominent source of power—our sun. Solar energy is provided daily, free of charge, if we can find a way to make it work for us. Photovoltaic cells are familiar on calculators and landscape lighting. Large cells have been used to provide electrical power for homes and office buildings. So far, applications for generating electrical power for large-scale distribution have not been economical. In the future, improvements in the efficiency of the cell and reductions in production costs may provide a clean, reliable, and truly renewable source of electric power.

Solar cells are not the only means of harnessing the sun. Solar chimneys use sunlight to heat air so that it rises through a giant chimney. The moving air drives turbines that generate an electric current. An Australian company is currently planning a plant that will use a chimney to produce 200 megawatts of solar thermal power, providing electricity to around 200,000 Australian households.

The ultimate new technology for energy could come from the same reaction that powers the sun. In nuclear fusion, hydrogen atoms combine to form helium atoms, releasing large amounts of energy in the process. There are significant challenges to using nuclear fusion to provide power. The greatest is learning how to handle material at a temperature in excess of several million degrees. In 2006, a consortium of the European Union, India, Japan, China, Russia, South Korea, and the United States agreed to jointly build and operate a research project, known as ITER. The goal of the project, which will cost about $9 billion, is to "demonstrate the scientific and technological feasibility of fusion energy for peaceful purposes."

Ff Fast Facts

According to the International Energy Agency (www.iea.org) in a statement released November 7, 2007, if governments around the world stick with their existing energy use policies, the world's energy needs will be well over 50 percent higher in 2030 than today. In the scenario presented, China and India together would account for 45 percent of the increase, as their energy use is set to more than double between 2005 and 2030.

Genetic Modification

Inside the cells of every living thing, genes carry instructions for cell activities. These genes, located in a section of the cell's DNA, have the codes that tell the cell how to make proteins that perform many functions inside the cell. Genetic modification (also called genetic engineering) is the direct manipulation of the genes inside a living cell. The purpose of genetic modification is to change the way the cell functions.

Genetic engineering is done by isolating a gene in an existing organism that has the codes for a desirable trait. For example, some plants produce chemicals that are toxic to insects. The production of these pesticides is directed by one or more genes in the plant cell. The second step of genetic modification is insertion of the isolated gene into the DNA of a different organism. If the gene that directs production of an insecticide is implanted into the DNA of a plant that does not naturally produce it, the new plant—and its offspring—will have the ability to produce the compound.

Genetic engineering has been used to produce plants that resist certain pests. It has also been used in agriculture to develop crop plants, such as corn, that are immune to a particular herbicide. When the herbicide is used in a cornfield, it kills weeds but not the resistant crop. Genetic modification has been widely used to produce plants with specific characteristics, such as resistance to a particular disease, tolerance for frost, or extended freshness.

In the future, genetic engineering may be widely used in medicine. Proteins are hard to build in chemical manufacturing processes, but living cells produce proteins constantly. It may be possible to produce many protein-based drugs using genetically modified organisms. Bacteria and yeasts are easily managed in laboratory settings. Domestic animals, such as cattle, sheep, and goats, produce milk, which is a protein-rich substance. Genetic engineering may lead to production of useful compounds that can be isolated from the milk of these animals.

 Fast Facts

People with diabetes do not correctly produce the protein insulin, which helps control sugar levels in the bloodstream. Until the 1980s, they had to use injections of insulin from cows or pigs. The first genetically engineered medicine was synthetic human insulin, approved by the United States Food and Drug Administration in 1982. Scientists used bacteria in which they inserted the directions for insulin manufacture. They were then able to use the bacteria to produce and harvest artificial insulin. Millions of people with diabetes now take human insulin produced by bacteria or yeast that is genetically identical to the insulin produced by human cells.

Researchers are currently looking for ways to modify genes within human cells. This may provide a new way to attack hereditary diseases. Gene therapy is a promising approach that offers a great deal of hope. The goal is to replace a defective gene with a healthy copy of the gene in order to correct a defective function in the cell. In 2000, a team of French scientists succeeded in curing young children suffering from a rare severe immune deficiency by inserting a therapeutic gene in the cells of their bone marrow.

Genetic engineering is just one of the many new technologies that you are likely to read about now and in the future.

Appendix A

Science Websites

There are many great books on science and its history where you can find more about the why and how of the universe, how we know the answers, and where we are going. Most people today also use the Internet as a major source of information.

It is impossible, in a book, to keep an up-to-date list of websites. Following are some of my favorites that I expect to still be around whenever you read this list. Think of this list as a starting point.

The usual caution about websites applies when you are seeking science information: be skeptical of any source whose reliability you cannot confirm. You can generally rely on established science magazines, government research labs, universities, and scientific and medical associations (see Appendix B) for reliable information. Beware inaccurate information if you don't know anything about the source or if it is sponsored by a person or an organization with a reason to be biased.

Encyclopedias

Online encyclopedias are often good places to start.

www.wikipedia.org   Many people regard Wikipedia as the ultimate encyclopedia. It is a good starting point for basic information about a topic, as well as finding definitions of unfamiliar terms. Unfortunately, its open-access format allows misinformation to find its way into articles too easily. I sometimes start with Wikipedia, but I never end there.

www.britannica.com   Reliable reference, but much of its information requires a subscription.

www.howstuffworks.com   Good articles in a question-and-answer format and lots of links for more information.

www.eol.org/index   The Encyclopedia of Life has a goal of providing reliable information on every species of living thing.

Finding Links

Your search engine can lead to many science sources. There are also some more specific ways to seek information sources.

Portal to U.S. government science sites: www.science.gov

Current research and news releases: www.eurekalert.org

Papers from government, scientific associations, and patents: www.scitopia.org/scitopia

Good starting point for information searches, including science (Internet Public Library): www.ipl.org

National Digital Science Library: http://nsdl.org

European Science Foundation: www.esf.org

Museums

Many museums have excellent websites, particularly in relation to their current displays. Here are two good museum sites and places to find hundreds more.

Links to a great list of natural history museums worldwide: www.lib.washington.edu/sla/natmus.html

Links to more science museums around the world: www.fi.edu/learn/hotlists/museums.php

Smithsonian Institute: www.si.edu

The Exploratorium: www.exploratorium.edu/index.html

Science News

Many established science magazines have online publications, and there are some useful and interesting web-only publications. These sources are a good place to look for recent research and news as well as good articles on any topic you can think of.

BBC Science: www.bbc.co.uk/sn

Discovery Channel: http://dsc.discovery.com

LiveScience: www.livescience.com

National Geographic: http://science.nationalgeographic.com/science

New Scientist: www.newscientist.com/home.ns

Nova (PBS): www.pbs.org/wgbh/nova

Popular Science: www.popsci.com

Science Daily: www.sciencedaily.com

Science News: www.sciencenews.org

ScienceNOW (AAAS): http://sciencenow.sciencemag.org

Scientific American: www.sciam.com

Seed Magazine: www.seedmagazine.com

Space.com: www.space.com

Technology Review: www.technologyreview.com

Questions for Scientists

There are many websites that allow you to address science questions to resident scientists. Most of them are sponsored by universities or research organizations. Here are a few places to find more science information in a question/answer format.

Cornell Center for Materials Research: www.ccmr.cornell.edu/education/ask

Newton BBS Ask a Scientist: www.newton.dep.anl.gov/askasci/env98.htm

Howard Hughes Medical Institute: http://askascientist.org

University of Wisconsin: http://whyfiles.org

Association for Astronomy Education: www.aae.org.uk/serv01.htm

Major U.S. Government Research Programs

Here are links to some of the major research efforts of the United States government. The portal listed previously, www.science.gov, will take you to many more. Other governments around the world also sponsor major research programs.

National Oceanic and Atmospheric Administration: www.noaa.gov

NASA: www.nasa.gov

Hubble Telescope: http://hubblesite.org

Earth Observatory: http://earthobservatory.nasa.gov

National Institutes for Health: http://medlineplus.gov

Food and Drug Administration: www.fda.gov/oc/science.html

Department of Energy: www.doe.gov

Universities

University websites are often great sources for science information. Check out the public information page, college of science, or specific departments to find out what research is currently going on. MIT is noted because its open courseware project puts materials from all of its courses online.

Major research universities (mostly United States): www.ura-hq.org/universities/index.html

Links to almost 8,000 universities and colleges: http://univ.cc

MIT Open Courseware: http://ocw.mit.edu/OcwWeb/web/home/home/index.htm

Appendix B

Professional Associations

Professional associations and societies are good places to find information about specific fields. Generally, these are organizations of people in a particular field. Many associations and societies, however, have good outreach and education programs to provide information to people who are interested in the topic but not professionals. There are thousands of such societies worldwide. This list includes some of the larger societies.

For a big list of science societies, covering every science field that you can imagine—and, probably, a few beyond that (for example, the Nigerian Dwarf Goat Association)—go to the Internet Public Library's list of Science and Technology Associations: www.ipl.org/div/aon/browse/sci00.00.00.

General Science

The American Association for the Advancement of Science: www.aaas.org

National Science Foundation: www.nsf.gov

Royal Society (United Kingdom): www.royalsociety.org

Physical Sciences

American Chemical Society: www.acs.org

American Institute of Physics: www.aip.org

American Physical Society: www.aps.org

Royal Society of Chemistry (United Kingdom): www.rsc.org

Biological Sciences

American Institute of Biological Sciences: www.aibs.org

Federation of American Societies for Experimental Biology: www.faseb.org

Earth and Space Sciences

American Astronomical Society: www.aas.org

American Geological Institute: www.agiweb.org

American Geophysical Union: www.agu.org

American Meteorological Society: www.ametsoc.org

International Astronomical Union: www.iau.org

Technology and Engineering

IEEE (originally the Institute of Electrical and Electronics Engineers): www.ieee.org

International Technology Education Association: www.iteaconnect.org

Index

X-Y-Z

x-rays, 268-269
    food irradiation and, 248

Yellowstone Park, fires in, 196
Yi Xing, 294

Zai-Lun, 294

Great gifts for *any* occasion!

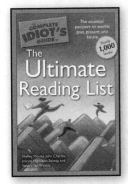

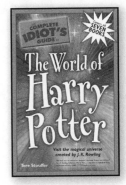

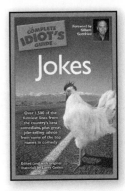

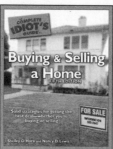
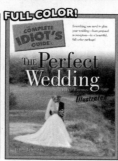

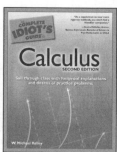

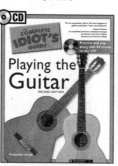